A SENSE
OF THE COSMOS

A Sense
of the Cosmos

Scientific Knowledge and Spiritual Truth

Jacob Needleman

MONKFISH BOOK PUBLISHING COMPANY
RHINEBECK, NEW YORK

First published in the USA by Routledge, Chapman and Hall, Inc. 1975

Book and cover design by Georgia Dent
Cover photo by NASA

Bulk purchase discounts, for educational or promotional purposes, are available. Contact the
publisher for more information.

First Monkfish edition

First impression

10 9 8 7 6 5 4 3 2 1

Monkfish Book Publishing Company
27 Lamoree Road
Rhinebeck, NY 12572
www.monkfishpublishing.com

For My Father

. . .Oh priest, consider mankind, how the phenomena of life and Nature are under the very eyes of them all; but they, in their puny selfishness, see only the few things of which they can make use. Very rare are those who seek the Cause for its own sake; very rare are those who allow themselves to be moved by those phenomena of periodicity, attraction and repulsion which are the manifestation of the *Ideas*. How many, do you think, are there of those who seek with their heart to divine the mystery which makes the waters rise to the sky and makes them descend again through the Nile to our Earth? . . . How many are those who, without arrogance, search for the power that moves and the law that is behind all this?

adapted from *Her-Bak: The Living Face of Ancient Egypt*
by Isha Schwaller de Lubicz

CONTENTS

PREFACE TO THE MONKFISH EDITION

For ten consecutive days in December of 1995, the NASA Hubble Space Telescope, orbiting far above the earth's atmosphere, pointed its lens toward what seemed an "uncluttered" portion of the sky in the constellation Ursa Major. Astronomers narrowed the focus of the telescope to a tiny speck of black sky about the size of a dime 75 feet away. The resulting image, assembled from over 300 separate exposures, appears on the cover of this book.

I was standing at a magazine rack in a Borders bookstore when I first saw this photograph on the front of the National Geographic. Opening the magazine and eagerly reading the explanation of the photograph, I was struck with wonder: a nearly microscopic point in an apparently empty patch of the night sky was here shown to be a window onto hundreds, thousands of stars, many certainly greater than our own sun, and, like our sun, pouring out unimaginable streams of life-creating energy onto who knows what planetary worlds and who knows what living beings that may have arisen upon these worlds. I remember standing there for a moment with my eyes closed, sensing the mingling of impersonal joy and yearning that everyone sometimes experiences looking up at a night sky strewn with millions of shining worlds.

I put the magazine back and started to walk away, but after two or three steps I stopped short. What had I actually seen? Something was not quite right. I turned around and went back to the magazine rack. My knees nearly gave way when I looked at the picture again. Could it really be? I opened the magazine again and this time very attentively read the explanation of the photograph: these were not stars at all, they were galaxies! Hundreds, thousands of galaxies never before known or seen inhabited that infinitesimal speck of "empty" sky, each galaxy itself containing billions of suns. I suddenly became very quiet inside.

I would like to think of the present book as an extended commentary on this image and the inner experience such images and discoveries can evoke. Every day, in almost all its branches, the revelations of modern science offer evidence that the universe, reality itself, is alive—alive beyond all imagining. All those who love science must know this truth in their bones, whatever may be the view officially sanctioned in the corridors of our universities and institutions of research. In any case, this is and always has been the view offered by the great spiritual traditions of the world, East and West, in all cultures and at all times previous to our own.

The very word "cosmos" signifies that the universe itself is a living organism, unimaginably vast in its extent and in the depth of its purposes and intelligence—and its beauty and, above all, in its goodness. And, according to these traditions, to know this universe, to know reality, it is necessary for a man or woman to perceive it with more than the intellect alone. It is necessary to perceive it with the unique source of perception by which beauty and goodness can be perceived—with the depth and subtlety of the power of feeling. The power of feeling—not the violence and chaos of what we usually know of as our emotional reactivity—the power of feeling must be joined to the genius of the intellect in order to know the nature of reality.

We cannot know, so the great spiritual traditions teach, with only one part of the human intelligence. To know with the intellect alone is to know beings, but not to know Being itself, which is where meaning resides. And this implies that the true scientist must himself or herself strive to bring together all the parts of oneself, must strive to become an ordered world in oneself as a prerequisite to seeing and knowing the order of the cosmos and the true nature of everything within the cosmos, all life, all elements, all laws and forces. Then one begins to understand that the great mechanism of the cosmos is an abstraction from, that is to say, an embedded aspect within, the far greater organism that is the universe, reality itself. Not only in our ourselves, in our own bodies, but in everything, everywhere, mechanism exists only as an aspect of organism. Mechanism is the instrument of organism; mechanism is the instrument of life, it is how life does things.

But really to know how life does things, it is necessary to know why life does things. And this cannot be known without the joining together in ourselves of feeling, instinct and thought. The greatness of modern science, as this book tries to show, is rooted in its courageous effort of reliance on what it considered the pure intellect as it was joined to and supported by a rediscovered respect for the bodily senses—which, in a larger meaning of the term, form part of the "instinctive" functions of the human psyche—as the source of knowledge. But in this revolutionary development of modern science, what was forgotten—for reasons having partly to do with the widespread cultural blunting and degradation of the meaning of faith in the religious institutions of the West—is that the heart, the power of profound feeling, is absolutely necessary in order both to be good and to see the good, to know the good that is an objective—yes, objective, attribute of the real world—out there. Losing this meaning of the human heart, losing the feeling component of knowing, science easily becomes scientism and easily leads to the belief that there is no objective value in the world. And this in turn leads to the moral relativism that has become the source of despair in our culture and especially in our younger generation. Of course, no less despairing, but not as obviously so, is what is simply the other side of the coin of moral relativism, namely, moral absolutism, the tyranny of the emotional reactivity masquerading as the human heart.

Here is the truly revolutionary aspect of this ancient vision: it is telling us that it is impossible for a human individual, for mankind, to have real knowledge without at the same time having virtue. When it is said—and it is said and seen by everyone now who has any eyes at all—that our knowledge has gone far beyond our morality, this is the same thing as to say that we need to rediscover how to join the attention of the heart to the powers of the mind and the perceptions of the senses. And this is to say, simply, that our being must catch up with our knowing. We must begin to confront a mysterious directive offered by two of the greatest minds of the twentieth century: Martin Heidegger and G.I. Gurdjieff. Each spoke in his own way—Heidegger as a philosopher, and Gurdjieff as an "awakener" – of the need to think deeply, to ponder, to contemplate the one ultimate question: the Being of beings. What can such words mean to us? And why should they be, how could they be, the most important question that our world has to face? They sound so abstract, maybe even meaningless, so removed from the flesh-and-blood problems of our world and everyday life.

But step outside one starry night. Go to a place where the "light pollution" of man-made cities is lessened. Go to a place out there and in here where our inventions of concepts and explanations no longer obscure the subtle intimations of higher truths within oneself. And look up at all those shining worlds.

What do you feel?

No. That is not the only question to ask oneself.

The question is: what do you know?

It is the same question.

Then observe your inner state. Could you hate? Could you be over-whelmed by envy or resentment? Could you dishonor any man or any woman? Is it not true that your wish to know more and more about the great world around you is now joined to the deep yearning to serve one's neighbor and whatever it is that is, for you and for me, God? Is it not true that no man or woman has ever committed a crime in the state of wonder? Is it not true that there is such a thing as sacred knowing? And can there be any real knowing, worthy of the name, that is not embedded in a sense of the sacred out there and in oneself? Does our world cry out for anything more funda-mental than this sense of the cosmos?

Jacob Needleman

March 2003

INTRODUCTION

Modern Man
Between Two Dreams

*I*t is necessary to think in a new way about science.
Once the hope of mankind, modern science has now become the
object of such mistrust and disappointment that it will probably never
again speak with its old authority. The crisis of ecology, the threat of atomic
war, and the disruption of the patterns of human life by advanced technology have all eroded what was once a general trust in the *goodness* of science.
And the appearance in our society of alien metaphysical systems, of "new
religions" sourced in the East, and of ideas and fragments of teachings emanating from ancient times have all contributed doubt about the *truth* of science. Even among scientists themselves there are signs of a metaphysical
rebellion. Modern men and women are searching for a new world view.

For several centuries Western civilization has operated under the
assumption that we can understand the universe without understanding ourselves. But having turned the available energy of our minds toward the external world, we now find ourselves more perplexed and anxious than ever in
front of a reality that simply will not yield to our hopes and desires. Our
technological achievements are great, but we see they have not brought

understanding.

Now—fitfully, and with great uncertainty—it seems we are being called back from the impulse to believe we can stride into nature with our mind pointed outward like an unsheathed sword. Both within and outside of the sciences a new sense of the unknown has appeared. The unknown is ourselves.

New teachings about man and his place in the cosmos are entering our culture from the Orient and the ancient worlds. These teachings from India, Tibet, China, and the Middle East; these ideas from the priests of Pharaonic Egypt and from the alchemists and mystics of antiquity now exist among us like the whisperings of another reality. And the discoveries of science about the organic interconnection of all things, from the atomic nucleus to the unfathomed psyche of man to the inconceivable entities of cosmic space, in a like manner invite us to something greater than the search for additional facts and explanations.

How will we respond to this invitation from the unknown? That is the question I wish to open in this book. I do not think it is a simple question, nor that the answer will necessarily be comforting. We may find that while something is now possible for us that has not been possible since the onset of the scientific revolution, something as well is demanded of us which is equally unprecedented. Some new effort within ourselves, some change of attitude so revolutionary and so uncompromising that it may very simply prove to be beyond us.

More and more one hears it said that the new religions from the East, with their "technologies of inwardness" and their encompassing metaphysical doctrines, are precisely what our epoch needs in order to humanize the thrust of modern science. The so-called "antiscience" movement among many of our young people is in part an expression of this feeling. The claim is that through the psychological development offered by the Eastern religions we can transform those moral and psychological flaws which have made our use of scientific discoveries so destructive. The goal of mastering ourselves to the extent that we have "mastered" nature now seems a real possibility to an increasing number of people.

Introduction: Modern Man Between Two Dreams

But perhaps we are only dreaming. What is required of us personally, privately, if we are not simply to replace a dream of outer progress with a dream of inner progress?

There is an oft-repeated saying of the ancient Greeks: "Whom the gods wish to destroy they first make mad." But have we understood this saying? What does it mean to be driven mad by the "gods"? The medieval alchemists said it more clearly: "Mother Nature sings a lullaby before she slays."

The lullaby of scientific progress, the dream of manipulating nature to suit our egoistic purposes, is ended. An increasing number of us, both scientists and nonscientists, pause rather longer and more quietly in front of the numerous breakthroughs in the sciences. And we are rather more sensitive to the ripple of new emotion that passes through us in front of the unknown. *In front of the unknown:* that means, when explanations break down and for a moment I am suspended between dreams. It is a moment of relative awakening.

But what is the new song that is now being heard by so many of us? Is it only another lullaby?

The premise of this book is that Western civilization as a whole now finds itself between dreams. In the true meaning of the word, it is a time of crisis—with all that implies of both extraordinary danger and opportunity. For there is nothing to guarantee that we will be able to remain long enough or deeply enough in front of the unknown, a psychological state which the great traditional Paths have always recognized as sacred. In that fleeting state between dreams, which is called "despair" in some Western teachings and "self-questioning" in Eastern traditions, an individual is said to be able to receive the truth, both about nature and his own possible role in the universal order. Throughout the ages, the hidden psychological methods of the ancient traditions have operated to guide people in that state between dreams, where we can begin the long and difficult work of self-investigation leading to transformation.

SCIENCE AND THE NEW MYSTICISM

Recently, I received a letter from a young physics professor asking me to advise him about changing his career in the direction of the study of spiritual traditions. "The most technologically advanced society in the world," he wrote, "is now the site of a rebirth of spiritual practice." In this development, he envisioned the possibility of a "humanization of science and technology and a transformation of religion." There followed five closely typed pages in which he carefully outlined all the issues involved and the themes he wished to explore. The Eastern religions, he said, had showed him that spiritual tradition itself can be viewed as a science and a technology, a kind of "internal science, in which the materials and apparatus are simply oneself." He went on to speak of the need to develop a deeper understanding of the symbolic modes of communication found in ancient traditionand he then put the question of whether modern science itself could be transformed into a spiritual path, what the Hindu tradition calls a "sadhana." He cited our contemporary visionary critics of modern science who urge "the return of science to its origins in Hermetic philosophy and alchemy or who foresee a new Pythagorean science." He proposed the creation of a discipline leading to a more direct and personal experience of scientific knowledge through which modern people could attain to a new consciousness of meaning in nature and human life.

Then, at the end of this long, carefully formulated letter, he added a hastily composed note in his own handwriting:

"I am searching for something for myself. I love science and I don't want to give it up. But it is not enough. I am searching for knowledge that is enough."

While reading the letter, I was mentally formulating replies to the various points he made. But the concluding words, which he had probably added just before putting the letter in an envelope, stopped me. At the last minute, he had dared to expose his real hope and it jolted me like the sudden appearance of another level of truth. It caused me to remember something that the writing of this book has brought me to again and again: *The real unknown is always an emotional unknown.* It is not merely a question of new, exciting facts about nature, or comprehensive new paradigms of explanation; nor is it a matter of new religions. The real question of the moment between dreams is whether we can bear the vibration of this new feeling of the unknown which carries with it the taste of a different quality of intelligence, but which at the same time utterly exposes all our illusions about ourselves. We *awaken to darkness.* This phrase has often come back to me during the writing of this book—along with the question: Do I fear the darkness more than I love the awakening?

Weeks went by before I mailed a reply to the letter. I tried to be academically proper and to give him information about careers in the fields of comparative religion and philosophy. But above all I wanted to respond to what he showed me at the end of his letter. I found myself settling for a few awkward words in a postscript:

Dear Dr. A—: Could it be that we are all looking in a wrong way for a knowledge greater than what science has offered? I wonder how you see this. Like you, I am surrounded by new religions, fragments of ancient spiritual traditions and new psychological methods for producing changes within ourselves. Like you, I hear people calling for a new synthesis of science and religion and I find myself dreaming of the transformation of human nature spoken of in the great teachings of the past. But, for myself, I begin to see how fragile and impermanent this wish for new knowledge actually is. I'm sure this is true of most of us, for there is very little in the conditions of modern life to support this wish and to help it ripen into a merciless self-interrogation. Is there, do you think, a way of approaching the truths of both modern science and the ancient traditions that fully takes into account this weakness in ourselves and all that is connected with it of impatience, fear and self-suggestibility?

Months went by during which I forgot about Dr. A. Then one morning I received another letter from him, informing me that he and a group of colleagues had organized a society called The New Pythagoreans. Their aim, he said, was to establish a community of scientists and spiritual leaders "to rekindle the vision of Pythagoras who brought to ancient Greece and to the Western world its first great fusion of spiritual discipline and the mathematical science of nature." Their project had already attracted the attention of several teachers from the Orient who were now living in America—a well-known swami, a Buddhist and a Sufi master. Their membership included physicists, biologists, a famous astronomer, a psychiatrist, a neurologist and representatives from the fields of philosophy, history and anthropology. Most with extremely solid names. Would I be interested in joining?

Attached to the letter was a brief, scholarly essay by the historian who had been named as one of the group's charter members. In it he outlined the background and subsequent influence of Pythagoreanism throughout the centuries, laying particular stress on the body of writings known as the *Hermetica*, which appeared in Egypt under the Roman Empire several hundred years after the birth of Christ. I had long been intrigued by the Hermetic tradition, though I had never thought of connecting it unequivocally with the origins of modern science. As in the reputed teachings of Pythagoras in ancient Greece, the Hermetic writings develop a concept of man as a mirror of the cosmic order, a microcosm. And throughout these ancient texts there are also hints of a personal discipline which is said to enable one to experience in himself the laws of a divinely ordered universe. The natural world is spoken of as the "book of God"; and the fully developed human being is understood to be the integration of all the purposes and energies of cosmic nature.

I had always surmised that the Hermetic teachings had once offered themselves as more than a mere system of belief, and that connected with them there had perhaps once been a key to the primordial science of awakening. But I had always assumed that this key was lost long ago, and that in more recent centuries all attempts to revive practical Hermeticism were mainly self-deceptions.

Not so, according to the article. The Hermetic tradition, so I read, had not died out after the Middle Ages (during which period it concealed itself under the symbolism of alchemy). In the fifteenth century the original Hermetic texts were translated for the first time into Latin and gradually entered the broad mainstream of European thought. "Eventually displacing the world view of Christianity, the teachings of the Hermeticists about the universe and the nature of man formed the basis of the scientific revolution. Modern science is actually the child of Western esotericism." The essay went on to list some of the ideas which were transmitted to Renaissance man through the Hermeticists: for example, the idea of an infinite universe and the idea of the human body as a great mechanism. Even the modern conception of the experimental method was traced back to the Hermetic teaching that "man has the power to know through direct experience all the secrets of the cosmos which are hidden in the microcosm."

All this struck an extremely responsive chord in me. The idea of the New Pythagoreans was that early on in the scientific era the exploration of nature became wrongly separated from the quest for self-knowledge. And this point corresponded almost exactly with one of the central issues of the present book—which at that time I had almost completed. As the reader will see, this book is directly concerned with the way certain ideas, which are meant to help us discover the truth for ourselves, become instead mere tranquilizers or even forms of psychological poison.

Yet although the program of the New Pythagoreans interested me very much, although it seemed a logical expression of the present hunger for a new kind of knowledge, something caused me to back away. It was not an intellectual judgment on my part, just as I cannot justify solely through intellectual reasons my hesitancy in front of most current attempts to correct the inadequacies of modern science through the enthusiastic adoption of new metaphysical ideas or spiritual techniques.

It is the same feeling—the impulse of plunging into a new dream—that I personally associate with that extraordinary period of transition between the Middle Ages and the modern, scientific era. I am not a professional historian, but it seems to me that then, as now, in the period we call "The Renaissance," Western man found himself between two dreams: behind him the dream of a Christianized world, before him the dream of the conquest of nature. In that period between dreams, some thing new entered into the life of our culture. Yet not all new things are automatically beneficial.

The expression of the teachings of Jesus which we call "medieval Christianity" was breaking down. The scholastic theologians had systematized

Christianity to such a point that little remained in it to call man into the state of total self-questioning. The passions, needs and aspirations of human beings could not be contained by scholastic thought; and the universe of Christian theology could no longer serve as the mansion within which general human life could proceed in all its vibration and color. At the same time, the extraordinary interaction of forces—spirituality, political power, the accumulation and abandonment of wealth and property—that had nourished the creative development of the monasteries was now dissolving. As for the organized Church, it won its battles too well. For centuries it had been a vital force precisely because it had constantly to rediscover its role in the interface between monastic asceticism on the one hand and the worldly claims of the secular state on the other. The life-giving interplay of these three forces—monastery, Church and state—came to an end as the Church both absorbed the monasteries into its organizational structure and also allowed itself to become too much of a secular power. Organized Christianity ceased to be an influence that could touch all sides of human existence— the life of the body, the emotions of family and social life and the aspirations of the mind.

Against this background, new ideas about man and the universe began to enter into the bloodstream of Western civilization. The result—at least, the result which we are concerned with here—was modern science.

Where did these ideas come from? And were they intentionally fed into the vortex of European life in the same way and from the same *kind* of source that had originally transmitted the teachings of Jesus into the life of the Western world? Or did these new ideas exert the mixed sort of influence they have eventually had because Western man could not bear to remain in the state between dreams?

I hope the above does not give a wrong impression. I do not claim to know where new, awakening ideas come from or how they need to be transmitted so as to serve as a positive influence on the life of a civilization. I do say, however, that this is a crucial question. And that it is not being spoken about very much at the present moment when so many people are turning to teachings that challenge the world view of science. We are so accustomed to believe that great truths need only to be put before us and they will have a beneficent effect. But I wonder if there is not something exceedingly naive in this assumption, some naive estimation of our unaided ability to *be* what we know, some failure to realize how swift and subtle is the passage from seeing the darkness to dreaming of light.

In any event, the great traditions make no such easy assumption about our ability to digest the truth. From one point of view, in fact, sacred tradition can even be defined as the science of transmitting truth by degrees so that it can enter correctly and harmoniously into the human psyche. To this end, a tradition both withholds and reveals at the same time. Transmission of truth is always understood in this way. There is always a "secret." Because there is always that in ourselves, which seeks only to believe and explain and to manipulate, rather than understand. We are calling that part of ourselves "the dreamer," but it has many names in the traditions, chief among which is "the ego." We shall have much occasion to speak of this in the following chapters as we explore the fate that overtakes living ideas when they fall into the hands of the dreamer.

There is always a secret, an unknown, because there are always these two sides of human nature which the traditions tell us must be kept separate and distinct. Each requires a different "food" in order to live and serve its purpose. Therefore, the great traditions speak to the ego in one way and to the other part of ourselves, which we have not yet named, in another way. At the same time, an authentic tradition offers itself as a guide by which to help us distinguish these two sides of ourselves so that we may recognize which part of ourself is active from moment to moment and so that we may see how we wrongly give to one part that which belongs to the other. I am speaking here about ideas which a teacher formulates differently according to the state of consciousness of those to whom he is speaking.

This book is therefore an effort to see modern science as an aspect of ourselves—much as a person would study his or her own mind, his or her own life, in order to learn precisely what sort of help is needed amid the colossal breakdown of one's world view, and in order that we might become sensitive to that help in its real and not illusory forms.

CHAPTER ONE

The Universe

THE UNIVERSE AS A TEACHING

Several years ago when I first started to write about the Eastern religions, which are now taking root in America, I could not understand why it was that every word I put down on paper seemed false, why every beginning ended in a lie. To write about our young people and their search, their experiences and struggles, that was more or less within my grasp. But when I turned to the towering spiritual systems of the Sufis or the Tibetans, for example, I very easily lost my way. Standing before these ancient teachings, which far surpass my understanding, I would often fall back on praising or comparing them.

Gradually, I began to see that great teachings enter the world according to an order and sequence that we are bound to find incomprehensible. But people are impatient to have a handle on what they do not understand. And so we fasten on one or another aspect of a system—an idea here, a method there—which satisfies our impatience. The result is that all we have before us is, so to say, a "cross-section" of the entire system. But obviously no number of static cross-sections can add up to the flowing structure of a living teaching.

Now I wish to write about the universe; and I wonder if the difficulties will be greater or less. Is the order of the universe any less organic than the order in the teaching of the Buddha or Jesus?

It may sound strange to compare the universe to a teaching, but we should realize that this is an absolutely fundamental question for us if we are to move toward a deeper understanding of our place in the cosmic order. It is not merely one author's personal brand of metaphor; quite the contrary. The point is that a spiritual teaching is true to the extent that its action in the world is in some strict sense an incarnation of fundamental cosmic laws.

Let us, therefore, entertain the possibility that we understand very little about what a sacred teaching or a universe really is.

Every great spiritual teaching speaks of itself in its own way as a mirror of cosmic reality. In the traditions of China the *Tao* is both the way to truth and the way things are. In Christianity the *Word* is both the teaching of Jesus Christ and the fundamental manifestation of God. In the Hindu tradition (including Buddhism) *Dharma* means both duty and the sustaining order of the universe. And in the Hebrew tradition *Torah* includes not only law in the sense of the teaching, but also law in the sense of the foundations of God's creation. A well-known passage in the Book of Proverbs expresses this idea without ambiguity. Wisdom is speaking:

> The Lord possessed me in the beginning of his way, before his works of old.
>
> I was set up from everlasting, from the beginning, or ever the earth was.
>
> When there were no depths, I was brought forth; when there were no fountains abounding with water...
>
> When he prepared the heavens, I was there: when he set a compass upon the face of the depth . . . when he gave the sea his decree, that the waters should not pass his commandment; when he appointed the foundations of the earth: then I was by him. . .
>
> Now therefore hearken unto me, 0 ye children: for blessed are they that keep my ways.
>
> Hear instruction, and be wise, and refuse it not.
>
> (Proverbs 8:22—33)

Wisdom thus speaks not only as the teaching (the instruction) but as the divine pattern of the cosmos.

How to think about this equation of the universe and a great teaching? It is tempting, for example, to see a teacher such as Mohammed or Bodhidharma, who brought Buddhism to China in the sixth century A.D., as the bearer of an extraordinary energy which is distributed to the world in the form of ideas, actions, events, schools, factions and the organized efforts of the community of followers. To compare this sort of pattern to a universe would require that we think of reality not in terms of things, but as a ladder of process, a great movement and exchange of energies. A teaching would then be a copy of this cosmic process on the scale of human time on earth. To receive such a teaching in ourselves, one's own life would have to become yet another copy of this process.

Thus, taking Christianity for a model, one might ask: What was the teaching of Jesus? Was it only what he said? Or does it not also include what he did and suffered? But does the teaching stop even there? A critic may claim that Jesus failed because Christian life has become what it has become. But is not the distortion, the crucifixion of the teaching, also, in a larger sense, part of the teaching itself? And if a man is to become a Christian, perhaps it is absolutely necessary that he witness the same process of distortion within himself. How else will he understand that it is in one's own thought and emotion that the "crucifixion," the distortion of the truth, really takes place?

Yet another line of speculation—again purely by way of opening this issue: Spiritual teaching is often spoken of as *indirect.* What is meant by this is that such a teaching does not act by persuasion, which is a form of compulsion and seduction, but rather by providing certain kinds of experiences. For a man who is searching for truth, these experiences are such that they cannot be assimilated only by a part of himself, the isolated intellect, for example. They require that a person receive them with the whole of himself.

Writing in the nineteenth century in a massive onslaught against the theologians and philosophers who wanted to make the Christian teachings accessible solely to the intellect, Soren Kierkegaard put the point as follows:

> The communication of results is an unnatural form of intercourse between man and man, in so far as every man is a spiritual being, for whom the truth consists in nothing else than the self-activity of personal appropriation, which the commun- ication of a result tends to prevent.

And then, comparing God to a teacher, he writes:

For no anonymous author can more cunningly conceal himself, no practitioner of the maieutic art [the art of the midwife] can more carefully withdraw himself from the direct relationship than God. He is in the creation, and present everywhere in it, but directly He is not there; and only when the individual turns to his inner self, and hence only in the inwardness of self-activity, does he have his attention aroused, and is enabled to see God.[1]

The prophets and spiritual innovators who have written of the universe as bearing the "signature of God" must surely have included something like the above in their thinking. Certainly, reality is as "silent" as any Zen master. And perhaps the only way for us to understand reality is through a more complete assimilation of the experiences which it presents us, both joyful and painful. *Yet the universe is so vast, our planet so small and our lives on it so inconsequential that a teaching is necessary in order for men to be exposed to the full range of events which take place in a cosmos.*

Pragmatism and Desire

We are trying to entertain the idea that the universe is like a great teaching so that we will be less afraid to question the picture of the universe that modern science gives us.

For most of us, the initial obstacle in this effort at skepticism is that science seems to *work* so well. A physicist I know once likened contemporary people, including himself and his fellow scientists, to a mob of savages so bedazzled by an interloper's tricks and baubles that they immediately make him into a god. It is really no laughing matter, this slavery to what we call the pragmatic criterion. We need to look at it more closely.

Imagine that a certain man comes upon a gun. He has never before seen or heard of such a thing. Nor, we must imagine, does he have any need to kill for food or defense. He picks up the gun, turns it around, knocks it against a stone. What is this object? He takes it home and experiments with it. To his delight he finds that when he holds it by the barrel he can crush things and break them better than with his wooden mallet. To him, the gun is a hammer. That is his idea, his theory, so to say, and his theory works.

When others ask him what that strange object is, he can prove his answer through the test of experience.

Why did this man not discover the proper nature of the gun? Because he did not ask of reality: How can I kill? And therefore reality never answered him or provided him with an instrument for killing. Here the limits of the pragmatic criterion are laid bare. When an idea or theory "works" it always does so relative to what we are asking of reality. If we have narrow intentions, our discoveries—no matter how ingenious—can never be bigger than our basic intentions.

This is what my physicist friend was speaking about. Rockets land on the moon, great bombs are exploded and certain diseases are treated. Such things so bedazzle us that we assume they are also answers to deep questions about reality. But the pragmatic successes of science need to be looked upon mainly as signs of the sort of questions the modern world is actually asking of reality.

If the man who came across the gun had been searching for a better way to kill, he would surely have discovered the real function of the gun. His intention would have matched the intention of the maker of the gun. And if someone else had tried to show him, by pragmatic proofs, how good a hammer it was, he would have laughed at him. So we may say that unless we are aware of our own aims, and unless we are sure that our aims correspond to the real purposes that exist in objects, then the pragmatic criterion is of little use as a key to knowledge.

We can imagine our man in search of a hammer puzzling for a moment over the bullets and gun's chamber, and then we can see him "improving" his discovery by fashioning a gun without bullets!

The truth of this simple example of the gun was brought home to me recently when I attended a seminar of medical scientists on the nature of gout. The speaker was a brilliant and well-known physiologist who carefully explained how for the most part, the disease was a result of the human body's tendency to overproduce uric acid. No other animal does this, he said, and therefore no other animal suffers from gout. "Here nature made a mistake," he said, "similar to the mistake she made with the vermiform appendix. There is no good reason for the body to produce so much uric acid."

I was amazed. What picture of the universe lies behind the statement that nature makes mistakes? But, far more important, what picture of ourselves lies behind the easy belief that we can discern such mistakes? I do not doubt that this man was on the track of some new way to alleviate the painful symptoms of gout, but that is not the point. What amazed me was that this was all being passed off as knowledge about the human body and its functions. Technology? Yes. The ingenious manipulation of a narrow spectrum of observed data? Certainly. And of benefit to those poor souls suffering from pain? Well, perhaps, let us grant him that, with qualifications to be elucidated later. But knowledge?

The audience of several hundred researchers and physicians kept a respectful silence and asked thoughtful, sympathetic questions. I can imagine a similar audience listening to a lecture by the man in our example as he points out the mistakes inherent in his newly discovered hammer.

The fact that we are bedazzled by the pragmatic successes of science shows us that when we pursue science our real intentions do not match what we sometimes claim to be searching for. We say we want knowledge about the universe, but we test our knowledge only by its logical consistency, its power to predict and its production of marvelous feats. Our real intention, therefore, is to satisfy our desires or allay our fears—desire for explanations, a sense of security, or material gain; fear of the unknown, death, pain and loneliness.

We must therefore recognize that there is a great difference between the wish for knowledge and the wish to satisfy desire, *which is the basis of pragmatism*. And that knowledge in the service of our ordinary desires may produce a very different picture of the universe than knowledge which is connected to other motives.

What, then, are we to say to those who compare modern theories of the origin and structure of the universe to the systems of cosmology that were part of the ancient teachings? Such comparisons are usually made in order to show how superior our present theories are and, occasionally, to point out the foreshadowings of modern science among these otherwise "prescientific" peoples. But was ancient man merely groping for what we believe we have found in our pragmatically testable hypotheses? Or did he mean something entirely different than we mean when he asked, "What is the origin and structure of the universe?"

A Conscious Universe

The scale of the universe is awesome. Our sun, which is more than a million times greater in volume than the earth, is, as everyone knows, only a tiny speck in the unimaginable vastness of the Milky Way. Hundreds of billions of such suns make up this galaxy, most of them far greater in size than our own. And the galaxy itself is but a tiny speck among countless billions of galaxies that occupy the cosmos that science perceives.

Each sun is an ocean of energy, one tiny fraction of which is enough to animate the life of our earth and everything that exists upon it.

Every second there pours forth from the Sun an amount of energy equal to four million tons of what we call matter. Since the planets of suns capture so little of this energy, all of outer space is in reality a plenum of force that is largely invisible to us, yet life-giving.

To set our minds reeling, it is enough to contemplate the bare distances that astronomy has measured. Light, traveling at 186,000 miles a second takes eight minutes to reach us from the sun—but four years from the nearest star, 27,000 years from the center of the Milky Way, and 800,000 years from the galaxy Andromeda. Yet Andromeda is now considered a member of what is called the local cluster of galaxies, beyond which lie countless stars and groupings of stars *thousands* of times more distant from us than Andromeda!

As with size, energy and distance, so with the reaches of time. Astronomers say the earth is some five billion years old, which means that the entire history of mankind, as we record it, is but a fraction of a second in the time scale of the earth.

It is no exaggeration to say that in this picture of the universe man is crushed. Within cosmic time he is less than the blinking of an eye. In size he is not even a speck. And his continued existence is solely at the mercy of such colossal dimensions of force that the most minor momentary change in these forces would be enough to obliterate instantly the very memory of human life.

Ancient man's scale of the universe is awesome, too, but in an entirely different way, and with entirely different consequences for the mind that contemplates it. Here humanity stands before a universe which exceeds it in quality as well as quantity. The spheres which encompass the earth in the cosmological schemes of antiquity and the Middle Ages represent levels of conscious energy and purpose which "surround" the earth much as the physiological function of an organ such as the heart "surrounds" or permeates each of the separate tissues which comprise it, or as the captain's destination "encompasses" or "pervades" the life and activity of every crewman on his ship.

In this understanding, the earth is inextricably enmeshed in a network of purposes, a ladder or hierarchy of intentions. To the ancient mind, this is the very meaning of the concept of organization and order. A cosmos—and, of course, *the* cosmos—is an organism, not in the sense of an unusually complicated industrial machine, but in the sense of a hierarchy of purposeful energies.

Here it is important to note that even in terms of physical astronomy ancient man did not use the word "earth" in the way we do. In his astonished and astonishing book, *Hamlet's Mill,* Giorgio de Santillana explains how misled we have been to think that the wise men of old actually thought the planet earth was *flat.* Cosmic phenomena were described, and their laws were expressed

> in the language, or terminology, of myth, where each key word was at least as "dark" as the equations and convergent series by means of which our modern scientific grammar is built up . . .

What was the "earth"?

> In the most general sense, the "earth" was the ideal plane laid through the ecliptic. The "dry earth," in a more specific sense, was the ideal

plane going through the celestial equator . . . the words "flat earth" do not correspond in any way to the fancies of the flat-earth fanatics who still infest the fringes of our society and who in the guise of a few preacher-friars made life miserable for Columbus. . . . (Moreover), the name of "true earth" (or of "the inhabited world") did not in any way denote our physical geoid for the archaics. It applies to the band of the zodiac, two dozen degrees right and left of the ecliptic, to the tracks of the "true inhabitants" of this world, namely, the planets.[2]

We have misunderstood these cosmological schemes of the past. What we call "geocentrism" was never meant to establish the earth merely as the spatial center of the great universe, but principally to communicate its place as an intersection of primary and secondary cosmic purposes and forces. The medieval mystic Meister Eckhart likens the earth to a station of cosmic reality through which there passes all the powers of Creation on their way to complete unfolding. "Earth . . . lies open to every celestial emanation. All the work and waste of heaven is caught midway in the sink of earth."[3]

In the Hermetic writings the hierarchical structure of the cosmos resembles that of an organism: cell in the service of tissue; tissue in the service of organ; organ in the service of the whole (governed by a supreme consciousness or intelligence). At each level of being there are "gods" or "angels" or, to use less uncomfortable language, "purposeful energies." From this point of view, the ancient spatial descriptions of the cosmos are meant to be understood symbolically.

Likewise, the word "sphere," used in describing the forces and purposes at different levels, is never meant merely to be taken literally. The very idea of the circularity of movement in "the heavens" can be understood to mean not only the encompassing nature of these progressively higher influences, but their *eternal* nature. The circle is, among many things, a symbol of that which "eternally recurs," that which is not subject to time and change as we know them.

Obviously, there is a great difference between contemplating a universe which exceeds us in size alone or in intricacy alone, and one which exceeds us in depth of purpose and intelligence. A universe of merely unimaginable size excludes us and crushes us. But a universe that is a manifestation of great consciousness and order *places* us, and therefore calls to us.

So much is obvious, for a conscious universe is the only reality that can include human consciousness. And only when I am completely included by something does the need arise for me to understand my relationship to it in all the aspects of my inner and outer life. Only a conscious universe is relevant to the whole of human life.

Undoubtedly, one contributing factor in our misunderstanding the cosmos of the ancient teachings is our habitual assumption that a conscious universe is somehow more comforting, a psychological crutch. Giorgio de Santillana also speaks to this in *Hamlet's Mill:*

> [Man] is unable to fit himself into the concepts of today's astrophysics short of schizophrenia. Modern man is facing the nonconceivable. Archaic man, however, kept a firm grip on the conceivable by framing within his cosmos an order of time and an eschatology that made sense to him and reserved a fate for his soul. Yet it was a prodigiously vast theory, with no concessions to merely human sentiments. It, too, dilated the mind beyond the bearable, although without destroying man's role in the cosmos. It was a ruthless metaphysics.[4]

"Ruthless" not in the sense of hostile to human hope, as many scholars have concluded by applying modern presuppositions to the interpretation of these ancient texts which speak of Nature as replete with "demons" and "darkness." The universe of the traditional teachings, such as Hinduism and Judaism, is "ruthless in that it is *ruthlessly responsive* to what human beings demand of it and of themselves. For whatever we expect from external reality reflects what we ask or fail to ask of ourself.

We must explore this thought further, for it can help us to see why the idea of a conscious universe appears to the modern mind as naïve. Science, as we know it, searches the universe for order and pattern. To pursue this search carefully, objectively, the scientist struggles to be free of his feelings, his inclinations to believe. He may follow hunches—what he calls "intuitions"—but in the final analysis he wishes for proofs that will compel the intellect, and only the intellect. The entire organization of modern science, the community of experimenters and researchers, the teaching of science in the schools, the training of specialists, is based on this ideal of proof that compels the mind.

Looked at in this way, we may conclude that the practice of modern science is based on a demand for human fragmentation, the division between thought and feeling. Searching for an outer unity, the scientist demands of himself an inner disunity. Perhaps "demands" is not the right word. We should simply say that in his practice the scientist endorses the division and inner fragmentation from which all of us suffer in our daily lives.

We now see why a conscious universe makes no sense to modern science. In the ancient teachings, higher mind or consciousness is never identified with thought associations, no matter how ingenious they may be. If these teachings speak of levels of reality higher than human thought, they are referring, among other things, to an order of intelligence that is inclusive of thought. Consciousness is another word for this power of active relationship or inclusion. Can the power to include ever be understood through a process of internal division and exclusion? Fascinated by the activity of thinking, and drawn to it to the extent of psychological lopsidedness, is it any wonder that we modern scientific people almost never directly experience in ourselves that quality of force which used to be called the Active Intellect, and which in the medieval cosmic scheme was symbolized by a great circle that included the entire created universe?

What Is Consciousness?

I realize that our task would be much easier if from now on we could be working with a precise definition of the word "consciousness." But it is important to stay flexible toward this question of the nature of consciousness. The word is used these days in so many different ways that out of sheer impatience one is tempted to single out one or another aspect of consciousness as its primary characteristic. The difficulty is compounded by the fact that our attitude toward knowledge of ourselves is like our attitude toward new discoveries about the external world. We so easily lose our balance when something extraordinary is discovered in science or when we come upon a new explanatory concept: Immediately the whole machinery of systematizing thought comes into play. Enthusiasm sets in, accompanied by a proliferation of utilitarian explanations, which then stand in the way of direct experiential encounters with surrounding life.

In a like manner, a new experience of one's self tempts us to believe we have discovered the sole direction for the development of consciousness, aliveness or—as it is sometimes called—presence. The same machinery of explanatory thought comes into play accompanied by pragmatic programs for "action." It is not only followers of the new religions who are victims of this tendency, taking fragments of traditional teachings which have led them to a new experience of themselves and building a subjective and missionary religion around them. This tendency in ourselves also accounts, as we shall see later, for much of the fragmentation of modern psychology, just as it accounts for the fragmentation in the natural sciences.

In order to warn us about this tendency in ourselves, the traditional teachings—as expressed in the *Bhagavad-Gita*, for example—make a fundamental distinction between *consciousness* on one hand and the *contents of consciousness* such as our perceptions of things, our sense of personal identity, our emotions and our thoughts in all their color and gradations on the other hand.

This ancient distinction has two crucial messages for us. On the one hand, it tells us that what we feel to be the best of ourselves as human beings is only part of a total structure containing layers of mind, feeling and sensation far more active, subtle and encompassing (like the cosmic spheres) than what we have settled for as our best. These layers are very numerous and need to be peeled back, as it were, or broken through one by one along the path of inner growth, until an individual touches in himself the fundamental intelligent force in the cosmos.

At the same time, this distinction also communicates that the search for consciousness is a constant necessity. It is telling us that anything in ourselves, no matter how fine, subtle or intelligent, no matter how virtuous or close to reality, no matter how still or violent—any action, any thought, any intuition or experience—immediately absorbs all our attention and automatically becomes transformed into contents around which gather all the opinions, feelings and distorted sensations that are the supports of our secondhand sense of identity. In short, we are told that the evolution of consciousness is always "vertical" to the constant stream of mental, emotional and sensory associations within the human organism, and comprehensive of them (somewhat like a "fourth dimension"). And, seen in this light, it is not really a question of concentric layers of awareness embedded like the skins of an onion within the self, but only one skin, one veil, that constantly forms regardless of the quality or intensity of the psychic field at any given moment.

Thus, in order to understand the nature of consciousness, I must here and now in this present moment be searching for a free quality of consciousness. All definitions, no matter how profound, are secondary. Even the formulations of ancient masters on this subject can be a diversion if I take them in a way that does not support the immediate personal effort to be aware of what is taking place in myself in the present moment.

In all that follows in this book, we shall continue to speak about levels of consciousness and intelligence within man and within the universe, for this idea is crucial in any attempt to reach a new understanding of science. But I wish, for the reader and for myself, that this more inner, personal meaning of the idea be constantly kept in mind.

Microcosmic Man

At this point it is necessary to introduce the idea of the microcosm, which will guide much of the thinking in this book.

Many statements of this ancient idea are so literal-minded as to make it seem incredible that people ever took it seriously. On the other hand, most contemporary attempts to make use of the idea of man as a universe transform it into something so metaphorical and commonplace as to make it equally incredible that anyone could ever doubt it.

Yet the mystery, that is to say the energy, of the idea of the microcosm has somehow survived even to this day. Of all the fragments that have come down to us from the ancient teachings it alone has resisted capture by either science or religion.

How to approach it? Were I to attempt a historical presentation I would have to summarize the metaphysical and psychological teachings of every great tradition that has exerted influence throughout recorded history. For, this idea in one or another form resides at the core of all traditions. In Judaism, Christianity and Islam, we are most familiar with its expression in the teaching that man is made in the image of God—God, the Creator and Preserver of the Universal Wholeness in all its gradations and levels. The traditions of India speak of the Divine, Cosmic Man whose dispersal into fragments constituted the creation of the world and whose re-collection is the sole essential task of human life. In Buddhism we find the doctrine that all the levels of being, from mineral up through the gods, are contained in Man—Man, the center and Man the all-embracing Void. The traditions of China revolve around the idea of the King, the Great Man who governs the parts of existence. One could go on with this listing— through the teachings of Egypt, sub-Saharan Africa, the American Indian; in Plato and in the Stoic philosophers; throughout the great tapestry of alchemy in all lands. The idea is everywhere.

Yet for all the force that this idea still contains, and despite the record of its presence in all cultures and times, it is obvious that the key to our understanding it is missing and needs to be rediscovered in our own experience. Otherwise, it could never have happened that of all the civilizations that we know, ours is the only one in which this idea not only does not occupy a central place, but is so far from the center of our thinking that when people—scientists or otherwise—make use of it, they do so as though they were coming across a "new image" or a "new slant."

Man is the universe in miniature—such is the bare statement of the idea of the microcosm. But as our conception of the universe is dictated to us by the scientific world view, the idea in this bald form adds nothing to our self-understanding. In this form, the idea tells us only that the same laws and substances that govern and constitute the stars also govern and constitute the human organism. But what kind of laws? And what kind of substances?

Our understanding of the microcosmos is thus severely constricted by our preconceptions about the cosmos. For, when we think about the universe, what do we picture to ourselves? Simply repeating that it is unimaginably vast and great has the inevitable effect of allowing our thought to come to rest, which is equivalent to the illusion of having grasped something about the Whole. The idea that the universe is in man therefore leaves us untouched.

But it is enough actively to imagine the little we know of what takes place on this small planet earth for us to glimpse the power in the idea of the microcosm. One thinks of both the long, slow formation of the continents and the instantaneous eruption of a volcano; the birth and death of species that inhabit the earth for millions of years compared with the minute life span of a single-celled organism; the constant movement everywhere of the winds and the stillness of rock and ice. There is the internal harmony of the ecosystem which is yet composed of conflict, mutual killing, fire and storm; there is gradual, subtle growth constantly in process in all things and the sudden destruction brought by earthquake, climactic change and disease; there are all possible movements upward and downward, collisions of fate everywhere at every moment.

But more than that, there are the laws that govern all these processes, the intelligence that adapts, reacts, creates and destroys within ever larger and more fundamental scales of intelligence and law. Is this intelligence, this all-penetrating hierarchy of purposeful law, something that is only of the earth? Or does it not pervade the whole of reality?

We must remember that these few examples are of processes and patterns among which we ourselves live and which we have more or less actually experienced. But if we now move in our imagination beyond the earth toward the complex life of planets and moons, and toward the sun, the stars and the galaxies . . .

At this point the question becomes serious: What does it mean that all this is in man? And not only in man as separate processes in all their variety, but as a *cosmos,* an ordered whole under the rule of a ladder of governing, lawful intelligence? The response seems clear: In whatever sense and whatever way all this is in man, it is not in *my* life or in *my* awareness. I, this individual person, pursue my life nowhere near an awareness in myself of this incredible spectrum of time, force and structure, not to mention the intelligence that governs it from without and within. This realization is the key to the idea of the microcosm. And it is precisely this key that is missing or unemphasized in almost every account of it that we may come upon: Man is a microcosm, but *I am not that man.*

With this key in hand, we may now admit the idea of microcosmic man into our thinking.

The "Parable" of Geocentrism

The Astronomer bowed. . . "Her-Bak, look west . . . now turn slowly to the east, watching the sky. What do you see?"

"As I turn I keep on seeing new stars."

"If you stand still all night what will you see?"

"I shall see the stars passing before me."

"What moves? The stars or yourself watching them?"

"If I stand still it must be the stars that move." HerBak paused. "Unless Earth turns as I have just done. Is this possible?"

The Astronomer smiled at his bewilderment. He gave the disciple time to think, then asked, "If the stars move, if Sun and Moon travel, why should Earth alone in the cosmos stand still? The idea repels you?"

"It would be strange," Her-Bak replied, "if one had to imagine that Earth, that seemed still in a shifting sky, was on the move. But everything I learn proves that my senses are subject to illusion . . . I don't dare deny that such movement takes place if you tell me it is so."[5]

The Astronomer watched Her-Bak benevolently. "I will make no statement," he said. "Your experience of illusion is enough to make you careful. What matters to Earth's inhabitants is that they should know of their vital connections with the sky. As to the movement of the stars, it is better to note what you see than to imagine what may deflect you from the real meaning. Then we will place ourselves at the center of the sky we are watching, where all star-movement is seen by reference to ourselves."

I am suggesting that it is we in the modern world who have been naive about the cosmos. We have rejected *geocentrism,* the idea that the earth is the center of the cosmos, for naive reasons. And consequently the great power in heliocentrism, the idea that the planets revolve around the sun, has become a destructive influence in the life of Western civilization.

There were indeed ancient systems of the universe in which the earth was said to move around the sun. Yet it is the geocentric conception which has publicly dominated the life of Western man up until the scientific revolution. In the above quotation there is the strong suggestion that the learned men of ancient Egypt held back the heliocentric understanding from the multitudes (the public conception was of the earth as a disc covered by the starry vault of heaven). Why? Why would the heliocentric system be kept secret?

Could it be that a certain psychological preparation, a level of existential maturity, is necessary before a human life can truly profit from the understanding that this planet is one of billions of dependent worlds revolving around great suns, themselves dependent, in a vast organic universe where all places are in movement and in which no physical center ever can exist?

Is it possible that geocentrism was originally meant as "a parable spoken to the multitudes?" And is the development of heliocentrism in modem times a flagrant example of what happens when a truth which can only be correctly valued by the "intelligence of the heart" is formulated and received in the intellect alone?

Lest we believe there is something self-glorifying about the "parable" of geocentrism, let us take a closer look at the situation of humanity in this view. For one thing, the image of the earth at the center of the universe communicates the idea of a vast convergence of forces upon our plane of reality. This is hardly a comforting thought, as we are rather accustomed to believe we recognize the major forces at work upon us. We tend to think that a powerful force, even if we do not understand it in itself, always makes itself known to us by its effects. But in ancient geocentrism the spheres or forces that surround the earth are both more powerful and more subtle than anything originating on the earth itself. Understood in this way, geocentrism humbles us and calls us to search for a finer understanding of the influences that shape our life and the life of our world.

It is therefore a great mistake to assume, as all modern writers have, that ancient geocentrism exaggerated mankind's importance in the scheme of things. For to be at the center meant in effect to be at the lowest rung in the ladder of influences that begin in the divinity which lies outside the spheres. Moreover, and this is essential, ancient geocentrism understood that the forces represented by the spheres acted not only upon the earth, but upon individual man. If individual human beings were unaware of the influences governing the life of their planet, this picture of the universe told them they were all the more unaware of the influences governing their everyday lives. And it is precisely this aspect of ancient geocentrism which modern scholars have either failed to appreciate or ignored completely.

It is geocentrism, without the idea of microcosmic man, which modern science rejected. But a purely external geocentrism was never the whole meaning of this idea in the ancient world. It is only we, who have lost the idea of the microcosm, who see it that way. But taken with the idea of the microcosm, geocentrism reminds us that objective reality contains many kinds of influences that can act upon us, that there is a scale of being to which man is born would he but search for it as diligently as he pursues the satisfactions of external life. It is we who imagine that geocentrism was merely a balm for the ego and a primitive astronomical theory. Because we, having lost the idea of the microcosm, separate scientific from existential ideas, we imagine that this separation is what ancient man was grappling for when on the contrary it was precisely what he was struggling against.

The Face of Reality

Now, what of heliocentrism, the idea of a universe containing countless suns and dependent worlds, of which ours is but one? What was its meaning for man in the ancient teachings?

Though at first it may seem unrelated, the following report may be a great help in thinking about this question. The author of these very personal observations is P. D. Ouspensky, best known as an expositor of the teachings of the great Russian master G. I. Gurdjieff.[6] The life of Ouspensky is a singular example of the search for an ancient knowledge that has been lost behind the contemporary forms of religion and science. As a young man, around the turn of the century, Ouspensky traveled throughout Asia and the Near East gathering bits and pieces of what he sensed was a vast and all-encompassing system which he was convinced existed on the earth somewhere, somehow. "I felt that there was a dead wall everywhere, even in mathematics, and I used to say at the time that professors were killing science in the same way as priests were killing religion." Before meeting Gurdjieff, he experimented on his own—in a time when such things were utterly unknown to the Western world—with yogic methods, drugs and altered states of consciousness. Yet nothing seemed to answer the deep presentiment, which he had felt from his earliest childhood, that in mankind's distant past there existed a great science of awakening that had been handed down from master to pupil throughout the ages. While working on a newspaper in the turbulent, prerevolutionary Moscow of 1915, he first met Gurdjieff, about whom he writes, "He and his ideas produced a very great impression on me. Very soon I realized that he had found many things for which I had been looking in India. I realized that I had met with a completely new system of thought surpassing all I knew before." From that point on, until his death in 1947, Ouspensky devoted his life to studying and transmitting the teaching of Gurdjieff.

These are his impressions standing in front of the Sphinx:

Yellowish-grey sand. Deep blue sky. In the distance the triangle of the Pyramid of Khephren, and just before me this strange, great face with its gaze directed into the distance.

I used often to go to Gizeh from Cairo, sit down on the sand before the Sphinx, look at it and try to understand it, understand the idea of the artists who created it. And on each and every occasion I experienced the same fear and terror of annihilation. I was swallowed up in its glance, a glance that spoke of mysteries beyond our power of comprehension.

The Sphinx lies on the Gizeh plateau, where the great pyramids stand, and where there are many other monuments, already discovered and still to be discovered, and a number of tombs of different epochs. The Sphinx lies in a hollow, above the level of which only its head, neck and part of its back project.

By whom, when and why the Sphinx was erected—of this nothing is known. Present-day archaeology takes the Sphinx to be prehistoric. This means that even for the most ancient of the ancient Egyptians, those of the first dynasties six to seven thousand years before the birth of Christ, the Sphinx was the same riddle as it is for us today.

From the stone tablet, inscribed with drawings and hieroglyphs, found between the paws of the Sphinx, it was once surmised that the figure represented the image of the Egyptian god Harmakuti, "The Sun on the Horizon." But it has long been agreed that this is an altogether unsatisfactory interpretation and that the inscription probably refers to the occasion of some partial restoration made comparatively recently.

As a matter of fact, the Sphinx is older than historical Egypt, older than her gods, older than the pyramids, which, in their turn, are much older than is thought.

The Sphinx is indisputably one of the most remarkable, if not the most remarkable, of the world's works of art. I know nothing that it would be possible to put side by side with it. It belongs indeed to quite another art than the art we know. Beings such as ourselves could not create a Sphinx. Nor can our culture create anything like it. The Sphinx appears unmistakably to be a relic of knowledge far greater than ours.

There is a tradition or theory that the Sphinx is a great, complex hieroglyph, or a book in stone, which contains the whole totality of ancient knowledge, and reveals itself to the man who can read this strange cipher which is embodied in the forms, correlations and measurements of the different parts of the Sphinx. This is the famous riddle of the Sphinx, which from the most ancient times so many wise men have attempted to solve.

Previously, when reading about the Sphinx, it had seemed to me that it would be necessary to approach it with the full equipment of a knowledge different from ours, with some new form of perception, some special kind of mathematics, and that without these aids it would be impossible to discover anything in it.

But when I saw the Sphinx for myself, I felt something in it that I had never read and never heard of, something that at once placed it for me among the most enigmatic and at the same time fundamental problems of life and the world.

The face of the Sphinx strikes one with wonder at the first glance. To begin with, it is quite a *modern* face. With the exception of the head ornament there is nothing of "ancient history" about it. For some reason I had feared that there would be. I had thought that the Sphinx would have a very "alien" face. But this is not the case. Its face is simple and understandable. It is only the way that it looks that is strange. The face is a good deal disfigured. But if you move away a little and look for a long time at the Sphinx, it is as if a kind of veil falls from its face, the triangles of the head ornament behind the ears become invisible, and before you there emerges clearly a complete and undamaged face with eyes which look over and beyond you into the unknown distance.

I remember sitting on the sand in front of the Sphinx—on the spot from which the second pyramid in the distance makes an exact triangle behind the Sphinx—and trying to understand, to read its glance. At first I saw only that the Sphinx looked beyond me into the distance. But soon I began to have a kind of vague, then a growing, uneasiness. Another moment, and I felt that the Sphinx was not seeing me, and not only was it not seeing, it could not see me, and not because I was too small in comparison with it or too insignificant in comparison with the profundity of wisdom it contained and guarded. Not at all. That would have been natural and comprehensible. The sense of annihilation and the terror of vanishing came from feeling myself in some way too transient for the Sphinx to be able to notice me. I felt that not only did these fleeting moments or hours which I could pass before it not exist for it, but that if I could stay under its gaze from birth to death, the whole of my life would flash by so swiftly for it that it could not notice me. Its glance was fixed on something else. It was the glance of a being who thinks in centuries and millenniums. I did not exist and could not exist for it. And I could not answer my own question—do I exist for myself? Do I, indeed, exist in any sort of sense, in any sort of relation? And in this thought, in this feeling, under this strange glance, there was an icy coldness. We are so accustomed to feel that we are, that we exist. Yet all at once, here, I felt that I did not exist, that there was no I, that I could not be so much as perceived.

And the Sphinx before me looked into the distance, beyond me, and its face seemed to reflect something that it saw, something which I could neither see nor understand.

Eternity! This word flashed into my consciousness and went through me with a sort of cold shudder. All ideas about time, about things, about life were becoming confused. I felt that in these moments, in which I stood before the Sphinx, it lived through all the events and happenings of thousands of years—and that on the other hand centuries passed for it like moments. How this could be I did not understand. But I felt that my consciousness grasped the shadow of the exalted fantasy or clairvoyance of the artists who had created the Sphinx. I touched the mystery but could neither define nor formulate it.

And only later, when all these impressions began to unite with those which I had formerly known and felt, the fringe of the curtain seemed to move, and I felt that I was beginning, slowly, to understand.[7]

Against the Literal Mind

I have cited this long passage not only to indicate, through the intuitions of a sophisticated observer, the possible reach of the ancient world's intellectual achievements, but to suggest that an exceptional state of consciousness may be necessary if we are to think at all intelligently about a nongeocentric cosmos. How many people pass before the Sphinx without intimating even a shadow of what Ouspensky was groping to express? And what, therefore, is inwardly required of us who wish to entertain in our minds the picture of a universe in which we are as dust?

In his highly regarded *A History of Science*, George Sarton makes the following comment in a chapter citing some of the astonishing aspects of the ancient Egyptian monuments:

> The Great Pyramids are so wonderful that some of the scholars who tried to penetrate their secrets became the victims of a mild form of insanity and ascribed to the ancient builders occult and metaphysical intentions and an esoteric knowledge the possession of which would have been even more marvelous than the mechanical and engineering ability that they certainly possessed.[8]

Substitute "the universe" for "the Great Pyramids" and you have a fairly accurate description of the attitude of many modern scientists toward the effort at understanding meaning in the cosmos. Like Professor Sarton, the modern astronomer tends to stand before a greater reality assuming that his state of consciousness and his intentions do not influence his perceptions and the relationships (or lack of them) which he finds before him. Is the universe any less organic and meaningful than the Sphinx or the Pyramids? Perhaps we should not laugh at people who take the Bible too literally until we recognize that as scientists we have exhibited the same sort of mentality toward the entire sweep of cosmic order.

At this point the comparison of the universe to a teaching can begin to help us. To be literal-minded about what is sacred means first of all, to trust one's first impressions, one's first mental associations. The presumption in this is enormous. When I take things "literally" I am presuming to be so in contact with myself, so whole in my power of response that I can instantly receive what is being communicated. My subsequent thought may become quite intricate and sophisticated, as it often becomes among scientists and biblical scholars, but the fact remains that all my complex and ingenious interpretations are resting on one mere split second of very partial receptivity. Have I ever *directly* observed the way thought influences perception and the way emotional associations influence thought? Can I discriminate between the deeper and more shallow reception of an idea or an impression? If I do not know myself in this direct way then I do not know the instrument by which data from the world are received by us, and then what can I possibly know about the universe? Certainly this is part of the reason why meditation (or contemplation) understood as the work of directly studying one's mind and feelings, was never separated from the study of nature in the ancient traditions.

There is nothing "mystical" about this. The literal mind is a mind out of contact with the whole human organism, a partial mind which trusts itself in its isolation from the very functions which make possible a fuller receptivity to reality. Quite as though I were in a laboratory equipped with numerous refined instruments, but chose instead to examine everything with a cheap pocket glass.

The literal mind is both wrongly active and wrongly passive. By using the former word, I wish to say that it is a hasty mind, compelled by fear or craving to affirm its own habits and associations. Such a mind, which believes things are as they appear to it, is an aspect of what is known as the *ego* in the traditional teachings. This ego is constantly hunting for ways to affirm itself; to persuade us that we *are* the ego.

33

An individual may be the slave of the literal mind and yet be what the world calls a "poet." But such a poet still weaves his symbols and interpretations around the perceptions of the literal mind. The sustained perception of objective meaning is, I believe, a very much rarer thing than we suppose. It requires a constant access to clear emotional intelligence, whereas much of what we call poetry is the imposition of subjective feelings upon a literal-minded perception of the world. Thus, both the universe and sacred writing are twisted by "interpretation," whether literal or so-called "metaphoric." So I hope the reader will not take what I am saying about the literal mind of modern science to be an endorsement of the "poetic" approach to reality.

Heliocentrism

To return to our main point, we may say that ordinary human thought, no matter how brilliant, is thought in the service of the ego. Sacred ideas, however, are a force against the ego. In the presence of a serious idea we become quiet; for a moment something appears in ourself that is bigger and more real than the ego. In that moment we see that we do not know what we are.

The history of the "warfare" between sacred teachings and human nature is in part the history of the struggle between ideas and thoughts. Great teachings, in recognition of this tendency of thought to serve what is peripheral and superficial in us, speak on many levels simultaneously. It is not that we are asked to deny the surface of sacred communications; it is that we must not remain frozen to it. And it is the literal mind, supported by logical systems or "poetic" interpretations, that keeps us fastened to the surface of both scripture and the world around us.

We take the universe literally when our thoughts about it draw us completely outside of ourselves, alienating us from our own depths. When, therefore, the sacred idea of heliocentrism is met by thought that is insulated from real feeling it serves to drive us crazy: to a despair at the meaninglessness of things, to grotesque assumptions about unperfected man's superiority. Is there any doubt that the modern scientific view of man's place in the universe is just such an expression of madness? We have converted a sacred idea into a thought which completely ignores what is most essential in the nongeocentrism of Buddhism, Hinduism and the cosmological systems of ancient Egypt and the Pythagoreans, namely, the idea of levels of power and intelligence in the cosmos, the objective symbolism of the Sun as a source not only of perceptible light and force, but of illumination and life corresponding to a central fire within man himself which, were he to come in touch with it, would gradually transform him.

Heliocentrism is the sacred or inner meaning of geocentrism. Which is to say that the real sense of heliocentrism cannot be grasped through the isolated intellect, but must be experienced in the play of inner and outer forces that influence one's own life. It is only as one ignores the idea of microcosmic man and its existential import that geocentrism and heliocentrism appear to contradict each other.

This, apparently, was what happened to both parties in the famous dispute between the Catholic Church and Galileo in the seventeenth century. Rightfully considered by many to be the father of the scientific revolution, the great Italian scientist was formally condemned as a heretic by the Roman Inquisition in the year 1633 for teaching the "Pythagorean" doctrine of heliocentrism. In modern times this trial and condemnation of Galileo has been regarded, along with the trial of Socrates in ancient Athens, as a classic example of the search for truth being crushed by dogmatism.

History does indeed show that on the whole the appointed representatives of the Church were ignorant, blindly dogmatic, frightened, and some of them even nefarious in their struggle against the heliocentric system of Galileo. But the main point here is that the Church seemed to have lost every clear sense of the relationship between the outer and inner cosmos. Lacking this understanding, it made the dispute into one concerning two opposing views of the external universe. It thus contributed heavily to the strange conviction so prevalent in the modern world that scientific observation and theory can actually threaten the sort of knowledge gained through spiritual discipline.

But when the dispute is seen from this point of view, Galileo himself loses some of his heroic stature. Courageous though he was in his effort to stand by his way of investigating the cosmos, it may now be asked whether he was only substituting one literalminded approach for another.

It is true that the Copernican-Galilean-Newtonian era gave the world a renewed sense of law in the universe. Surely every serious student of modern science knows those moments when the intellectual grasp of a lawful pattern in nature frees him from his own subjective perceptions of what is before him, embroiled as these perceptions are in the tormented machinations of the ego. This brief release from ordinary thought, which is a foretaste of inner freedom, occurs when the mind is actually touched by a relatively objective idea.

Why then did the modern world forget that so much of the value of apprehending scientific law lies just in this quality of direct self-knowledge which such apprehending brings? How did we not see that if a general law of nature is objective it is also a law of our own nature? And that the deepened quality of our own experience in that moment is also an attribute of the universe? From this understanding an individual may surely sense that he lives his life in ignorance of the levels of conscious order that exist in the cosmos and which are hidden in himself.

Whatever the reasons may be for this forgetting, the fact is that after Galileo scientists began to pride themselves on not asking *why* things are the way they are, but only *how*. Because the Church had lost sight of the connection between cosmic and psychological purpose in the universe, the whole idea of purposes in nature fell into disrepute. The Church had become unable to do what is an essential task of all religions to do: to communicate the purposes of existence in such a way that an individual human being can hope to experience them both in himself and in the cosmos. The Christian view of the universe was reduced instead to dogma, in the sense of beliefs held without any method of verifying them for oneself. Modern science therefore rejected the wrong thing: It separated itself from the idea of purpose in the universe, when it should have rejected only the Church's wrong relationship to that idea. The Church had come to read the book of nature through hard and dead categories. Science, while beginning as a search for a new way to experience the meaning of that book, soon ended by counting commas.

Gradually, but inexorably, the desire to manipulate nature moved to center stage. Pragmatism was born, and the purpose of knowledge came to be the satisfaction of desire rather than the growth of consciousness. Theories were judged by predictive power or aesthetic appeal; appearances were judged by further appearances. The heliocentric theory was true because it gratified the isolated intellect, the ego and the desires. A sacred idea, which in ancient Egypt was given to people only after they had experienced the sense of their place and task in a hierarchically structured conscious universe—an objective idea which required the preparation of recognizing the limitations of one's own unperfected inner state—was taken by the ego, rendered external and became a negative influence upon civilization.

NOTES

1 Sören Kierkegaard, *Concluding Unscientific Postscript*, Princeton, New Jersey, Princeton University Press, 1969, pp. 217—18.

2 Giorgio de Santillana and Hertha von Dechend, *Hamlet's Mill*, Boston, Gambit, Inc., 1969, pp. 58—62.

3 Meister Eckhart, *Sermons and Collations* XX, translated by C. de B. Evans, London, Wat kins. Quoted in J. H. Reyner, *The Universe of Relationships*, Vincent Stuart, London, 1960, p. 21.

4 *Hamlet's Mill*, op. cit. pp. 5—6.

5 Isha Schwaller de Lubicz, *Her-Bak: Disciple*, London, Hodder and Stough-ton, 1967, p. 101.

6 See P. D. Ouspensky, *In Search of the Miraculous*, New York, Harcourt, Brace and World, 1949.

7 P. D. Ouspensky, *A New Model of the Universe*, New York, Alfred A. Knopf, 1961.

8 Ibid.

CHAPTER TWO

The Science of Medicine
and the Fear of Death

> Medical science is full of mysteries and
> must be studied like the words of Christ.
>
> PARACELSUS

*I*n 1543 Nicolaus Copernicus published his epoch-making work, *On the Revolution of Celestial Orbs,* the first modern, mathematical demonstration of the heliocentric theory. In the same year a remarkable young Belgian physician, Andreas Vesalius, published an anatomical text that was to have equally profound repercussions on Western man's understanding of himself. Called *On the Fabric of the Human Body (De humani corporis fabrica),* it contained a series of magnificent illustrations, unsurpassed to this day, of the skeletal, muscular, vascular and neural structure of the body as a whole. Never before had the human body been represented with such accuracy, exactly as it appears to the eye of the anatomist. For the first time, the body was seen—as it is still seen today—as a natural mechanism.

According to the eminent historian of medicine, Charles Singer, Vesalius' book constitutes the foundation of modern medical science. "The work of Copernicus removed the Earth from the center of the universe; that of Vesalius revealed the real structure of man's body."[1] Vesalius looked with his own eyes at what others had been content to accept on authority about the structure of the principal organs, the placement of bones, the distribution of the blood vessels and the nervous system, therewith calling into question the teachings of Galen, the Greek physician of late antiquity whose system of medical knowledge dominated Europe for fourteen hundred years.

The Science of Medicine and the Fear of Death

In the preface of De *Fabrica,* addressed to Emperor Charles V (for whom he was court physician), Vesalius writes of the delight which his monarch will surely take learning about "the temporary dwelling-place and instrument of the immortal soul":

> For this [human body] in many particulars exhibits a marvelous correspondence with the universe, and for that reason was by them of old not inappropriately styled "a little universe." . . . I am of the opinion that out of the whole Apolline discipline, and indeed out of the whole philosophy of nature, nothing could be fashioned more pleasing or more acceptable to Your Majesty than an account from which we learn of the body and of the mind and furthermore of a certain divine power consisting of the harmony of both, in sum, of ourselves, whom to know is man's proper study.[2]

At first slowly and against considerable resistance, but then with rapid acceleration, the Western world accepted the concept of the body as a great machine obeying the same natural laws as the stars and the stones. One disease after another seemed to yield to the revolutionized science of medicine. And the ancient association between *self-knowledge and medical science,* now only fleetingly mentioned by Vesalius, finally faded into the background and was eventually forgotten. The body was a mechanism like the universe itself. The sole purpose of medicine was to obtain information about the workings of the organism and the causes of disease in order to prolong life and relieve suffering. Moreover, as the concept of the universe itself became more and more materialistic, more and more a cosmos devoid of a governing intelligence, the body was also so regarded. The human body was still seen as a reflection of the universe, but of a universe that no longer contained God (or Self). In time, almost all the ancient and medieval medical concepts, for example, the doctrine of the humors, were cast aside as superstitious relics.

Concerning the notion of bodily humors, at the root of this doctrine was the idea of several basic organizing forces within the human organism, whose interaction mirrored the play of forces in the Creation itself: forces of fixation and coherence (connected to the elemental principle called "earth"); of dispersion and malleability (connected to the element symbolized by the manifest properties of "water"); of rarification into finer and subtler material ("air"); and, finally, of transformation upward into a new and higher unity ("fire"). Thus, in the medieval system of medicine, as in those of the Orient, the human body was implicated in the great ladder of being reaching from God down to mineral and upward again. This view of the body could not last long after the triumph of the modern scientific view of the universe which brought with it no means of verifying the existence of such forces either in the cosmos or in man.

Almost five hundred years before Vesalius, the Islamic philosopher Al-Ghazzali wrote "Man has been truly termed a microcosm,' or little world in himself, and the structure of his body should be studied not only by those who wish to become doctors, but by those who wish to attain to a more intimate knowledge of God." This double nature of medical science—an instrument for healing and a means for remembering the existence of a higher level of governing intelligence in the universe—was perhaps in Vesalius' mind as well, for he too was attracted to the Pythagorean and Platonic scheme of reality. But fifteenth-century Europe was not eleventh-century Baghdad or Isfahan. In the medieval Islamic empire the greatest philosophers and scientists were more often than not followers of the religion of Islam, though no outsider can say which of them was also a Sufi as well. The outstanding example, for our purposes, is the towering figure of Avicenna who not only wrote some of the most famous visionary tracts of the time, but was also the single most important medical mind of the Western world during the period between the eleventh and the fifteenth centuries. His principal medical text, *The Canon of Medicine,* assimilated ancient Greek scientific rationalism and empiricism into the context of the spiritual metaphysics of Islam.

I cannot here offer a detailed discussion of Avicenna's text, but we may note one significant aspect of the opening chapters of the *Canon,* where Avicenna is expositing the fundamental principles of his system. At the end of each of these chapters—dealing, for example, with the concept of bodily humors, or the elements (earth, water, air and fire) or the natural faculties of man—Avicenna adds a phrase to indicate that while he has said all that can be said for the practicing physician, there are many deeper aspects to these subjects which are relevant only to students of "esoteric philosophy." Perhaps Avicenna wished to remind his readers and himself that teachings which are meant to awaken the spirit cannot be mixed and diluted with formulations intended for pragmatic and theoretical applications.

In the writings of Vesalius and Copernicus (and many of the other Renaissance figures who laid the foundations of the scientific revolution) we find on the contrary little sense of a separation between awakening ideas and ideas that can serve the immediate material well-being of humanity. As to the causes for this, it would be idle of me to try to speculate beyond what I have suggested in the Introduction about the general breakdown of the Christian religion and the dying out of the few monasteries that may have sheltered the disciplines of the Christian Path. But whatever the causes, the fact before us is that the great ideas of ancient times are presented in the Renaissance in ways that stimulate the sense of man's *power* to explain nature, master nature and to act in ways reflective of the power of the Creator. The idea of the microcosm as a *possibility* to which man may aspire is subtly and inexorably altered to be a description of man as he already *is*. No greater distortion of that idea could be imagined.

As the modern era progressed and the sciences rapidly acquired their astonishing explanatory and manipulatory powers, the science of medicine kept pace. The history of modern medicine quite exactly reflects the development of the modern era's troubled relationship to nature, that is, our stance as a "conqueror." But here it is the human body that takes the place of the external world of nature. Scientific medicine approached the body exactly as physics, chemistry and astronomy approached the physical universe.

In the last decades of the nineteenth century and in the first half of the twentieth century the technological and theoretical triumphs of the physical sciences were matched most spectacularly in medicine by the doctrine of specific etiology, the theory that all diseases are caused by a single specific agent such as a microbe. Working from this theory and guided by the brilliant discoveries of such men as Pasteur (who developed the notion of bacterial infection); Robert Koch (who laid the foundations for the science of bacteriology) and Paul Ehrlich (who "targeted" the syphilis microbe and pioneered the special discipline of immunology), medical science had seemingly solved the problem of human disease, at least in principle. For any particular disease, it was a matter of identifying the particular microorganism that caused it and then devising a drug to destroy that microorganism. Some dramatic successes with diseases such as yellow fever, malaria and diptheria reinforced the theory of specific etiology and the vision of a coming medical utopia.

But this has proved to be a mirage, apparently for the same reasons that the whole vision of the technological conquest of nature has proved to be a mirage. Both the harmony of the body and the harmony of the environment seem to be far subtler and more powerful than we have realized. Whether it is antibiotics or pesticides, excessive surgical intervention or the plundering of forests, the overuse of drugs for symptomatic relief or the massive diversion of natural energy resources, neither the body nor the environment is so easily manipulated for the sake of satisfying our desires or allaying our fears. More and more we are beginning to see the extent to which disease is but one expression of mankind's total pattern of living.

And thus the ancient association between religion and medicine can no longer be lightly dismissed—of course, always remembering that by religion we do not mean contemporary religion, but a system of values, human relationships and a view of humanity's place in the cosmos which, taken together, serve as a foundation for the way an entire civilization lives, feels, thinks and acts. The "epidemics" of the latter part of the twentieth century are surely a sufficient indication of the truth of this claim. I am referring not only to the as yet unmeasured effects of generally increased radiation and hydrocarbons in our environment, but also the numerous modern diseases such as AIDS, vascular disorders, certain forms of cancer, drug addiction, boredom, neurosis and hypertension, not to mention the variety of ill-defined microbial ailments such as bronchitis and sinusitis which, in the words of one research physician, "do not take life, but just ruin it."[3] Not to mention the widespread physical effects of our own brands of psychological tension. Surely, future historians of medicine will perceive the connection between such contemporary diseases and the twentieth-century's contradictory attitudes about nature, pleasure, work, sexuality, the family, education, ideas and death—just as present historians of medicine are beginning to see that it was largely the changes in the social order dictated by the values of nineteenth-century industrialism that created the conditions in which certain microorganisms could act as pathogenic factors of epidemic proportions.

I cannot pretend to know whether or not the science of medicine during the Middle Ages was as ineffectual and unenlightened as modern historians say it was. But we can surmise that under the influence of a vigorous Christian *religion* the issue was: How can human beings struggle against bodily illness and death without forgetting *all* that is possible for man? How to prevent the human fear of pain and death from cutting off all avenues leading to the search for oneself? Surely, the perennial association of religion and medicine in traditional societies may be understood in this light—particularly in those civilizations, such as Pharaonic Egypt and ancient China, where an astonishingly well-developed system of practical, scientific medicine was blended with spiritual metaphysics, "magical" formulae and religious rites and symbols.

Of course, a great deal more is involved in assessing the relative merits of modern medical technology than these tiny hints can suggest. But our aim here is mainly to see that there are two approaches to knowing the human body-one which has lead individual men and women closer to the search for self-knowledge and one which serves the natural desire of the human species to resist death and physical disease. I have suggested that the ancient association of religion and medicine may have been originally intended as a means of distinguishing these two approaches to the body and preventing the latter from obscuring the former. And I am now suggesting that this distinction and the need for it was obliterated with the breakdown of the Christian religion and the birth of modern medicine.

The result is twofold:

(1) The sense of wonder evoked by our contemplation of the workings of the human body now strengthens our egoism instead of directing our attention to laws of movement and operation that are the properties of a

higher consciousness in action; and **(2)** fundamental truths about the organization of the human body are absorbed and distorted by our fear of death, and thereby serve to aggravate that fear to the extent that the fact of our own death has become completely unimaginable to us.

Let us look more closely at these factors in the hope of arriving at a wider perspective from which to view our ideas about the human body and our inability to feel the factuality of death.

Wonder and Its Counterfeit

When I was quite young, I remember anxiously saving my money in order to send away for a book called, I think, *Wonders of Nature*. Actually, it contained only a great many ordinary pictures of animals, insects, trees and the human body. But the main idea the book communicated was the absolute miracle that these living things should exist at all. For some reason, this made an exceedingly strong impression on me and I remember looking at everything in nature quite differently because of it. A miracle that this tree or that animal should exist. A miracle that it continues to exist. A miracle that it does not die right on the spot. To put it abstractly, the book persuaded me that there is nothing in the core of reality to support life, that life is a strange freak of chance, unsanctioned by the great universe. The author's intention was probably to enhance young people's sense of wonder. But in my case the book produced a purely mental wonder which I think actually cut me off from that more organic emotion which the ancient Greeks identified as the prime motive in the search for wisdom.

The impression stayed with me that there is no reason at all for living things to keep on living. As I grew up and continued to read scientific books, this attitude spread more and more toward myself, toward my own body. One summer, when I was a premedical student, I accepted a job assisting in a hospital autopsy room. My task was to prepare all the cadavers for the doctors and to help them while they made their examinations. I remember the mixture of fascination and fear as I sawed through the skulls of these dead bodies, removing and handling the brain, tying off the arteries, turning it around in the light to study the convolutions, and finally slicing the spongy mass like a ripe fruit in order to prepare a section for the histology lab. I can still smell the dust of the skull, and I remember the impression of animal life in the human guts which I had to remove and throw into the incinerator where they gave off an aroma like that of broiling food.

I was cutting two, sometimes four corpses a day. At the same time, I was reading every medical book I could get hold of. From everything I understood in these medical texts, and from what I saw every day with my own eyes and heard from the doctors who patiently answered all my questions, I found myself facing an intolerable contradiction. The human body, my human body, was so marvelously constructed, so complexly unified, so resilient, so intelligently adapted in the interrelationship of its parts with respect to the world around it, that it was incredible it should ever die at all. But for the same reasons it was equally incredible that it should continue to live more than a fraction of a second in a universe whose basic make-up was so alien to it.

The more I studied the workings of the body, the more my "wonder" increased. But how much of it was that imitation wonder, based on fear, which tempts us to become explainers and "conquerors" rather than conscious participants in a structure that includes the whole of ourselves? I wanted to know the body so that someday I could explain it and perhaps protect myself from it. Similarly, although the ideal of knowing God through knowing the human machine pervaded the revolution in medical science in the sixteenth and seventeenth centuries, what was the underlying motive? To study the body was to see how God did things — in fact, I remember thinking almost in these very words during one period when I was enraptured by the workings of the human kidney. As if to say: "So that's how God does things." But at the back of our minds, not in words, is the feeling "Some day, when *I* become God, that's how *I* will do things."

In short, there is a kind of wonder which accompanies the perception of a difference of levels in the universe and in ourselves. But it seems we then all too easily dream of striding beyond this difference of levels, rather than seeking to allow the higher level to enter our minds as a kind of guide to our unknown selves. That is why, it seems to me, we need the discipline of a

Path, and the ideas which it brings in ways that interfere with the explanations of the egoistic cerebral automatism. Since the Renaissance we have been dreaming we are great enough and intelligent enough to be taught directly by the universe and by the human body. Like every other emotion, our sense of wonder has become mixed with fear and egoism.

Perhaps there were once teachings, perhaps there still are, which can interfere with this habitual passage from wonder to egoism when we contemplate the workings of nature and the human body. We shall return to this question in the following chapters. But now we ask, what has been the result of our approaching the body solely as conquerors? What relationship to ourselves has been the legacy of modern science's attitude toward the body?

The War of the Ego Against the Body

Why do we have bodies? What is the body for?

Modern psychologists have blamed the teachings of Christianity for our inability to live with this question, taking St. Paul as the arch villain. But Paul's condemnation of the "sins of the flesh" *(sarx)* is the condemnation not of the body, but of a wrong and self-deceiving relationship to the body, the submission of that in ourselves which is designed to rule to that in ourselves which is designed to serve. The body as such is for Paul a "servant," and good.

It is impossible to say exactly when the attitude of hatred and fear of the body became associated with Christianity. Even the medieval monks who practiced "mortification of the *flesh*" did not necessarily understand the body itself to be the enemy. For many of them the enemy was the tendency to form a fixed image of oneself based on experiences of physical pleasure and pain. This is the essence of the idea of sin for both Christianity and Judaism: the attraction to a false or incomplete picture of one's own possibilities. In "sin," man sells himself short, as is illustrated in the biblical story of Esau, who forfeited his birthright because of an empty stomach.

As in other traditional teachings, Christian asceticism was in part a struggle to break down our psychological dependence on the accidents of external bodily sensations. Between this understanding and what we now call

46

"puritanism" there obviously lies a wide gap. The true ascetic struggles against the power of his false sense of self, the ego which is formed out of a distorted attraction to pleasure and an excessive fear of its absence. According to Evelyn Underhill's sensitive and reliable study of medieval Christian mysticism, the aim of bodily asceticism was the death (mortification) not of the body, but of selfhood (the ego) in its narrow individualistic sense. The process of mortification, she writes,

> is necessary, not because the legitimate exercise of the senses is opposed to Divine Reality, but because those senses have usurped a place beyond their station; become the focus of energy, steadily drained the vitality of the self. "The dogs have taken the children's meat." - . . It is thanks to this wrong distribution of energy . . . that "in order to approach the Absolute, mystics must withdraw from everything, even themselves."[4]

What we may call the "puritanical" struggle, on the other hand, is a struggle not against a false sense of myself, but against pleasure as such. The "puritan" attempts to affirm "my" power, the power of the ego, over the body.

The ascetic struggled against the body in order to destroy an unreal picture of himself. What he sought was not the destruction of the body, but a natural relationship to it, a relationship corresponding to the structure of universal Creation itself, in which consciousness and intelligence determine the operation of the physical world.

The history of Christian monasticism attests to the perennial difficulty in preventing the struggle with the "flesh" from becoming a battle between the ego and the natural demands of the body. The great monastic reformers, such as St. Benedict in the sixth century, seem often to have had this problem in mind when determining the degree and quality of physical discipline that was necessary at any given time. For example, Benedict's aim was, in his words, "to form a school of divine servitude, in which, we trust, nothing too heavy or rigorous will be established."[5] To him the extreme physical austerities of the earlier desert monks often tended to degenerate into a form of self-conquest, rather than self-surrender. In any event, the version of puritanism against which the modern world has so strongly reacted surely represents a distortion of the original Christian teachings.

Yet despite all our attempts to flee from "puritanism," the forces which generated it remain dominant in our society, due in no small measure to the influence of medical science. We have already seen how the modern world took the ancient idea of heliocentrism in such a literal way that it became a pathological influence, tempting us to pit ourselves against the entire universe. That is the sort of stance which I am now calling "puritanism." "Puritanism" I define as the egoistic fear of nature, the fantastic but nonetheless violent struggle to maintain a narrow sense of personal identity through manipulating the forces of nature, especially as they operate through the human body.

The seventeenth-century philosopher Rene Descartes first gave voice to this modern approach to nature. Starting with his famous statement, "I think, therefore I am," he first of all identified the self with the associations of thought. He then separated this "self" off from surrounding nature, including the body. And the body he then understood to be devoid of consciousness, purpose and the inherent power of life. We now know the implications of this "puritanism," though only in terms of the disruption of our biological environment due to our facing the external world solely as explainers and conquerors. But what of the inner biological environment, the human organism? Do we not also face the body in the same way as we have faced nature out there?

It is commonly thought that the contemporary world has swung from puritanism to hedonism—to the pursuit of pleasure rather than the denial of pleasure. But these are two sides of one coin. Both the hedonist and the puritan face the body in the condition of fear; the puritan fears gratification while the hedonist fears the absence of gratification. Both derive their sense of identity through conflict with the natural rhythms of the organism; both are manipulators, at war with what is.

It is therefore a mistake to think that modern medical science, including psychiatry, offers a relationship to the body significantly different from that of puritanism. The substitution of the love of pleasure for the hatred of pleasure means nothing here. It is all puritanism in a larger sense. The conflict between ego and nature remains.

We need to remember that by the word "ego" something much less hypothetical is meant here than what psychoanalysis intends by the term. The ego is the deeply ingrained picture I have of myself: of my qualities, my rights, my powers and my possible destiny. The conflict between the ego and the body is the effort to maintain a secure sense of myself on the basis of what I attempt to make happen in the body or to the body. *And this conflict must obviously have its effect on the bodily functions themselves.*

The Body as a Field of Force

There is nothing in present-day medical theory to suggest that this warfare between the ego and the body is itself the cause of disease. Most of the breakthroughs of modern medical science were the result of understanding the body and its illnesses solely in terms of impersonal mechanisms. Like the universe, the body was regarded as an automatism. The parallels here are so striking that one might well equate the ego's relationship to the body with modern man's relationship to the universe. The body which is the universe of the false "I" is a universe stripped of consciousness and purpose, devoid of living interrelationships, *impersonal* in the particularly modern sense of the word. And just as modern man sets off as a "conqueror" of his impersonal universe, the individual ego likewise takes the bodily automatism as a field for "conquest." The pragmatic criterion of success is physical pleasure and the avoidance of pain and death.

We have already seen that the ancient teachings regarded cosmic law as the "signature of God," the necessary pattern formed by the movement of divine, creative energy. The alchemists of the Middle Ages applied this same understanding to the processes within the human body. For the authentic alchemists, the truly human struggle is to open oneself to the *full* range of energies circulating through the body. The transformation of metals symbolized the process, or "work," by which the forces of nature within the body are brought into harmony as servants of consciousness. The enemy is the ego, and its repetitive preoccupation with a narrow spectrum of these energies, its recurring, habitual postures of fear and desire. The ego cannot master nor even perceive these basic energies of the human organism; no more than man in an ordinary state of consciousness can master or perceive the fundamental forces in the cosmos. Now, the ego is my superficial sense of myself, of personal identity. These central energies of the human organism, therefore, which are neither obedient nor even visible to the ego, are *impersonal.* This is the main meaning of the impersonality of the natural and cosmic world in the ancient teachings. Reality is not amenable to *my* desires, *my* concepts, *my* efforts at control. But this by no means implies that consciousness, intention and will are not properties of reality. It means only that such consciousness, intention and will are inaccessible to the egoistic personality of undeveloped man.

The modern scientific view of the body and of nature has retained something of this ancient idea of impersonal cosmic law. But it has added to it something which completely changes the human stance in favor of belief in the ego. To the modern mind, impersonal law means law devoid of conscious purpose, therefore something *less* aware (though physically more powerful) than the human personality. For the ancient mind, impersonal law refers to a higher level of intelligence and purpose than the ego can know, thus something *greater* than the human personality.

Disease as a Punishment

It thus takes great psychological maturity to see the human body and the universe in terms of impersonal law. Recognizing this, the great traditional teachings often held back the more scientific, impersonal formulations about reality that were to be found in the disciplines of the Path. Instead, religion presented its ideas about nature and the human body in forms and symbols which could, to some extent, penetrate the initial screen of egoistic thought and attract some of the emotional energy which was bound to the ego. This is one useful way to think about the distinction between the esoteric or hidden aspect of a great teaching and its exoteric or popular aspect, which makes every possible adjustment to undeveloped man's patterns of thinking.

It need only be added that to be impersonal about reality does not mean to be without feeling. It does, however, demand a degree of mastery over feelings that are egoistic, such as self-pity, hatred and most forms of fear.

Like the geocentric picture of the universe, the idea that *disease is a punishment* strikes us as naive, superstitious and perhaps even dangerous. Like the geocentric view, it smacks of an era when, so we believe, humanity overestimated its own importance in the scheme of things and failed to understand the rigorously lawful nature of reality.

Certainly, modern medical science has removed any shadow of guilt from the disease process. We do recognize that certain diseases are in part brought on by ourselves and the way we choose to live, a very banal truth which does not invoke the question of our relationship to a greater reality. Psychosomatic medicine is supposed to deal with such illnesses. In short, we believe we have learned to be impersonal not only about the functions of the body, but about the psychological processes which influence them.

But have we? Before laughing at the idea of disease as punishment, we should be sure we are not making the same mistake we made in thinking about geocentrism and in hastily labeling it as naively "prescientific." We may find that we have been entertaining only the crudest and least intelligent sense of this idea as well. To think impersonally about my body may require the same sort of psychological preparation that is required in order to be impersonal about the cosmos.

Punishment, then, for what? We have already seen that in the traditional teachings the good of human life is much more of an inward thing than we have taken it to be. It concerns the establishment of a new order *within* ourselves, a psychodynamic integrity so complete and spanning such a range of forces as to mirror the great cosmos itself. If this is our true possibility, then clearly anything that moves us toward that goal is good, and anything that impedes us or distracts us from it is evil.

When we read that the body is good, but the flesh is evil, we are being told that the body is designed to serve this quest for inner transformation. The idea that disease is a punishment is therefore part of the larger idea that the human body is designed to respond to influences that are much higher than we may imagine. To be subject to a higher influence means to serve the purposes of a force that is more intelligent and conscious. To speak of disease as a "punishment" is a way of saying that the human body languishes when cut off from the conscious energies that it is built to contain.

Why Do We Have Bodies?

How obvious it now becomes that we need the idea of microcosmic man to understand the significance of the ancient teachings about health and disease. If the body is subject to conscious influences from without, and if man is a potential microcosm, it means that the body is potentially subject to *the same higher influences coming from within myself.*

Now, if the body of *microcosmic man* is the servant of interior universal forces, then what can the body be in man's unperfected state? The answer is: *It must become an instrument of his search.* From being an ally of the ego, the body must become a servant in the struggle for consciousness and understanding. But this can never happen if the body is *prematurely* regarded in an impersonal manner—as a machine for the production of pleasure and egoistic affirmation.

The impersonal view of the body fostered by modern medicine has very much the same consequences as the impersonal view of the cosmos fostered by modern science in general. In both cases man persuades himself that he is free to choose his manner of relationship to a reality that, in fact, includes him and influences him at every turn. But in order for an individual to relate freely there must be something in him that is free to relate, something separate from or independent of the body. This independent "something" is precisely that consciousness which is spoken of in the perennial teachings, and preciously that which unperfected man lacks.

We have seen that with regard to our understanding of our place in the universal order, this self-deception leads to a sort of cosmic paranoia. Man must first be free enough from his egoistic emotions if he is to make real use of the idea that the cosmos is impersonal. Otherwise, he equates impersonality with lifelessness, causal order with indifference.

Similarly, with regard to the body, the illusion that we have free attitudes toward the body merely disguises the fact that our ordinary thought and feelings are servants of the ego, preoccupied with physical and psychological pleasure and pain as the index of life's meaning. It is as much a presumption to believe we are able to grasp the idea of the body as a machine as it is to believe we are able to bear the idea of an infinite universe. We cannot bear either truth until we have a more intelligent grasp of reality as a conscious organization of laws and purposeful energies to which we may become openly receptive. And then the whole idea of causal order undergoes a profound correction; the mechanical laws of both the universe and the body are understood not as shackles that bind us, but as expressions of a higher will that is potentially our own.

Paracelsus, the remarkable physician and *alchemist* of the sixteenth century, expresses the idea of disease as punishment in the following way:

> Health and sickness are granted by God: nothing comes from man ... You should divide the diseases of men . . . into those which arise in a natural way, and those which come upon us as God's scourges. For take good note of it: God has sent us some diseases as a punishment, as a warning, as a sign by which we know that our affairs are naught, that our knowledge rests upon no firm foundation, and that the truth is not known to us, but that we are inadequate and fragmentary in all ways, and that no ability or knowledge is ours.[6]

It need only be added that Paracelsus is here speaking to and about the ego.

Summing up, the traditional teachings about the body may serve to warn us that it is not so simple to regard the body as a machine. In order to do this, we must seriously reflect upon the full range of universal energies which may act upon and within the human organism, a range that extends far beyond what modern science is able to acknowledge. Otherwise, the attitude of objectivity toward the body remains merely a screen for ignorance and self-deception about the place which unperfected man occupies in the cosmic order.

DEATH

Like most people, I have one or two memories about death which stand apart in my mind, and against which all philosophical ideas seem like straw.

One such memory is of an afternoon many years ago when I was six years old. I had just hurried home from school, leaving my mother somewhere down the street chatting with a neighbor. The moment I entered the house I felt a silence that made me shudder slightly. I remember experiencing for a split second a state that I would now call seriousness. It was definitely not fear; it was much more impersonal than that.

In the room at the opposite end of the hallway I saw my grandmother on the floor, her head slumped against the wall. I very clearly remember walking up to her; and as I walked, suddenly, two different consciousnesses existed in me, each with different sets of perceptions, different thoughts, different feelings. One "consciousness" knew exactly what had happened and accepted it as though nothing essential had changed or could change. The other consciousness was not sure and did not want to have anything to do with the first. I remember two very distinct sets of perceptions occurring as my hand calmly reached out and touched her face, which was cold and ashen.

I bolted out of the house and ran down the street to fetch my mother, shouting that Grandmother had *fallen asleep on the floor*. At that moment I saw that I wished to be deceived—or, to be exact, that one consciousness had absolutely no relationship to the perceptions of the other consciousness.

We rushed back to the house and the doctor was called, but by that time the awareness of two separate consciousnesses had disappeared. I experienced many strong feelings, which are still fairly clear in my memory, but I was already completely settled in the consciousness that had wished to be deceived. Gradually, I "adjusted" to what was going on.

What was that other consciousness which instantly understood everything objectively and impersonally? It was certainly worlds apart from what any modern psychologist has hypothesized as the "unconscious." For one thing, it was a consciousness utterly without desire or fear. And for another, it was connected with a general bodily sensation that had nothing to do with pleasure or pain as we usually experience them. Thinking back on it now, and on other similar moments, the impression is of contact with a deeply *central truth* about man.

Perhaps because I was only a child, I forgot about it as soon as it went away. And even the few later experiences of that sort never preoccupied me once they passed. Not even when, as a university student, I was reading mystical literature of all sorts did it ever occur to me to connect these experiences to what such books called "higher consciousness." There was no sense of "melting" into anything; quite the contrary. There was nothing which even a young man wanting to be someone special could call "divine bliss." That child of six had no thought associations by which to either like or dislike it. And in any case, he, that is, I, was so shaken by the immediate reality of death that any scientific or religious categories which might have been in my mind would probably have been obliterated anyway.

Several points arise out of this that are germane to the subject of this chapter.

First, it took the shock of the death of a loved one to bring even a child of six, who was relatively free of conceptualizations, into a state where impersonal perception was possible, if only for a few moments. A state, that is, where an impersonal consciousness of the world could exist. Here the word "impersonal" should be taken quite directly to mean a consciousness independent of what we ordinarily feel to be our personality. From this it follows that to be impersonal or objective has relatively little to do with what concepts one entertains, scientific or otherwise, but with quality or state of consciousness.

Second, the sort of experience I have described convinces me without any doubt that real objectivity is very far from being unemotional. On the contrary, in that moment there existed, for the short while that I was aware of it, the embryo of something I am compelled to call "love." I call it that simply because all my other feelings about my grandmother suddenly existed off to one side and were almost all colored by fear. Yet at the same time, blended with that impartial perception of the situation there was, without words or images, an acceptance, a taking in of the totality of the person. I shall have to leave it at that, because any other way of expressing this would have a sentimental tinge, and this emotion was absolutely without sentimentality.

Third, and most important, the consciousness in which we exist all the time and which contains all our thoughts, perceptions, sensations and feelings is completely cut off from that other consciousness which knows about death. No matter how intense our feelings, no matter how hard we try and how sincerely we wish to be truthful and know things for what they are, there is no way we can call into being a direct relationship between these two consciousnesses. We can do many things that will lead to the production in ourselves of strong emotions and important insights, but we have no way of coming directly in touch with that other consciousness.

I say this quite emphatically because, as clearly as I now see my hand before my face, I saw that I could not bear that other consciousness. And I am not speaking of being able to tolerate *what* it saw—it never even came to that. I could not bear that it existed; I saw every thought and-feeling swirling around in myself and how they were all connected together and all based on the assumption that they were all there was in myself and of myself. Like a colony of animals each feeding the other and none aware or able to be aware of the greater world in which they lived.

It therefore seems to me that the fact of death is not unlike the fact of an infinite universe. Death exists on a scale that is utterly incommensurate with the world which we perceive in our usual state of consciousness. We are accustomed to think of vast divergences of size, space and time when we reflect on the idea of the scale of things. But just as there are states in which we can for an instant sense the infinity of the cosmos, so there are states in which we can for an instant sense the finitude of ourselves. My own finitude is a fact as incommensurable with my ordinary reality as the fact of endless time or the idea of billions of galaxies.

In my professional career I have participated in many conferences with doctors, therapists and clergymen on the subject of death. Not once have any of us mentioned the awesomeness of death. We have all been eager to confess that death frightens us and leads us into self-deceptive denials of mortality. But never a hint of the fact of two consciousnesses in man totally unrelated to each other, only one of which can perceive reality on the same scale upon which death exists.

Death Is Awesome as the Universe Is Awesome

Death is awesome as the universe is awesome. Both are realities far beyond the scale of the ego. Now, the ancient teachings almost always gave man the geocentric universe to enable him to think without paranoia about his place and task in the cosmos. But what of death? How to begin to be toward this reality? What ideas can help me to think about death when I am utterly disconnected from that consciousness in myself which alone is able to *see* death as something concrete and factual?

This question is of truly surpassing importance. It cuts across the dualism of East and West, ancient and modern. Under the scale of the question of death, we are of one family with the builders of the Sphinx, the singers of the Vedas and the apostles of Christ, just as under the scale of the infinite universe we are of one family with the earth, the planets and the sun, all of which are almost as nothing in the vastness of the total Creation.

Under the scale of death, ideas of humanity's progress or degeneration are almost of trifling importance. The earth and sun will die; our eon will end, and—according to many teachings, including the teachings of modern science—the universe itself will sometime cease to exist. Of what ultimate significance is it, then, that modern conditions of life are less natural—less in tune with Reason—than were the conditions of life in ancient societies? We must accept that the Egyptians and the Aryans and Hebrews, as well as ourselves, needed constantly to struggle for even a moment's relationship with that second consciousness, not to mention the labors necessary to obey that consciousness in the activity of living.

Yet, for all that, there is one aspect of modern thinking that adds unnecessarily to man's difficulties in facing the factuality of death. I am referring again to the modern, materialistic view of a dead universe.

If one perceives the universe as a great system of consciousness, as a "teaching," the universal law of death can do more than merely frighten the

ego. How? Consider the way a great spiritual master, in order to liberate hi:
pupil, may undermine everything the disciple holds onto as a support for hi:
ego. Yet at the same time the teacher's presence before the pupil affirms—
without words or persuasion—something immeasurably greater which can·
not be grasped by the egoistic consciousness. This is *double communication*—
sometimes called "indirect communication"—corresponding to the double
nature of man. The second consciousness is addressed in such a way that the
ego is helped not to interfere.

In a similar way, the felt idea of a conscious universe can "permeate" the
perception of the law of death. In this way the ego is not simply tempted by
fear to affirm itself apart from the whole man. Without the idea of a conscious universe, even the merest glimpse of the cosmic scale of death leads
the ego into many human perversities, such as sentimentality, despair and
defiance. Sentimentality in violently held beliefs that comfort the ego;
despair in intellectual systems such as existentialism which glorify anxiety;
defiance in much of what we see of scientific technology.

Not even the merest edge of the mystery of death in life and life in death
is visible when we lose sight of the awesomeness of death. As the scale of
the universe was leveled by scientific metaphysics, so also the scale of death
was leveled. Until finally, as in very recent times, the scale of death is dimin-
ished to oneself. Now it is *my* death, the fact of which can only terrify the
ego into an insane "honesty" such as we find among our philosophers and
artists. Death preoccupies us without compelling us to feel our lack of
knowing who and what we are, and in what world we exist. Having been lost
in the illusion that we stand over against the whole universe, we now believe
we can come to grips with death merely by thinking about it, by finding the
right categories—religious, psychological or scientific—by which to explain
our relationship to it. But we have no relationship to death.

A New Question

What is needed if we are to make a sane approach to the reality of death,
or, to put it better, if we are to allow the reality of death to make us sane?
What can counter our present tendency to take death so personally that we

reduce cosmic law to the scale of our ego? What can make the fact that *I* am going to die and that you are going to die something that can awaken me to the search for a more fundamental reality in myself?

We have said that the idea of a conscious universe is needed alongside an intimation of the scale of death. But we know that such ideas cannot enter our being merely through our thinking about them (thus, Spinoza's saying that the wise do not think about death). No, *such ideas must enter our body.*

But how? Perhaps there are experiences which are instrumental and pre-paratory just as there are conceptual formulations which prepare the mind for great ideas. Could the experience of pain and disease make the fact of death concrete for us, and therefore an awakening force?

Pain, Ego and "Immortality"

I saw the problem for myself recently due to an ailment that kept me in more or less constant severe pain for several months. Although the pain was intense, it was limited to my legs and did not affect my vital functions. More-over, at no time was it possible to believe my life was threatened. Therefore, though racked by pain day and night, and though praying to every god and devil I knew that it end, I was able to make some clear observations about pain—observations which are now indelibly stamped in my mind. No one can ever tell me that it is possible to learn such things from books or merely from observing others.

What I saw was both ludicrous and terrifying. For the first two or three weeks, my life remained generally the same. Certain that the pain would soon pass, as it had in the past, I was "brave." I abjured pain-killing drugs. I complained and joked. I stayed away from doctors. I drank in sympathy from others. I, the ego, was rather pleased despite everything, and I actually admired myself for the way I was bearing up.

But the pain did not get better; it got worse, and eventually my compo-sure dissolved. I became frightened, and during the next two months I tried every doctor and healer I could find and every nostrum I could lay my hand on. They all failed, but each time I felt sure I was on the track of what was wrong and how to fix it—and each time panic set in when the timetable broke down.

Gradually, I began to notice something important out of the corner of my eye. I saw that my efforts were being expended not simply to stop the pain or to become well, *but to get back my habits.* I discovered that I loved my habits—my habits of eating and sleeping, sitting and walking, talking on the phone, meeting people, reading, laughing, shouting. When I took pain-killing drugs (which I began to do, in great amounts), I was relieved not simply because the pain lessened, *but because my habits returned.*

I learned that I am a being composed of habits, and that I depend on my habits to feel alive. *When I wish for immortality, I wish for the immortality of my habits.*

I do not want to appear naive. I had certainly always known how disturbing it is to have one's habitual ways interfered with for any reason. But I am speaking about something different here, something about human nature which I had never heard of before. When I was in pain, I was actually a freer human being. My habits no longer compelled me; I was no longer *lost* in them, I could see them and sense them as though they were children calling out to me not to abandon them. But I could not bear this freedom because there was one habit which not even the constant pain could dislodge: I am speaking of the habitual feeling *that I know what to do,* the sense one has in all circumstances day and night that I am doing the right thing, or at least the best thing possible. It is something much faster than thought and it is not usually formulated in words, but it pervades every action and movement of my life. And I think in this everyone is alike. Even in the most difficult situations when action is blocked on every side and I become limp, there is still the accompanying sense that this is what I must do. Even the most "passive" among us feel this beneath the surface of their lives. This feeling, *this sense of agency,* is something very small, but since it is I, it is also something very big. This habit is "myself." It is the ego.

Such as we are, therefore, not even physical pain can help us assimilate the factuality of death.

Toward a Constant Relationship to the Fact of Death

It is true that sooner or later life brings everyone occasions when death becomes a concrete fact. Upon the death of a loved one, or in a moment when my own life is suddenly in jeopardy it can happen that an extraordinary change takes place within myself. My knowledge of the fact of death is for an instant something that is alive in my body.

Many writers have attempted to describe such moments. The quality of time is transformed; a minute by the clock may contain so many subtle impressions that the whole sense of passing time is transcended. Experiences of people who have almost drowned or who have faced execution certainly fall into this category. One sees, as it is said, "one's whole life." *One sees one's whole self.* And one is amazed to know that it is also not one's whole self. There is something more, something one has never been in touch with, but which yet has a taste of intense familiarity. I have called it the second consciousness, but the name is not important.

There is a metaphysics and a psychology that emanate from such experiences, a view of reality and human nature. But conceptualizations about "higher dimensions" and the "immortal soul" are so quickly captured by the habitual associations of thought and emotional reaction that this metaphysics is usually wasted on us.

Therefore, the question arises: Must we rely only on rare accidental shocks to make death concrete? Spinoza's saying that the wise do not think about death may be understood in the light of this question. For surely the problem is not so much to resist thinking about death, but to understand that death cannot be faced in our everyday state of consciousness. The problem is how to recognize that thoughts about death do not make it concrete.

Then, are we condemned to go through life locked in daydreams and nightmares about death? Is there nothing we can do to voluntarily create an experience similar to death within the scale of our own individual existence, so as to discover something of the truth about death in our present state of consciousness? Behind this question there lies the strategy of self-knowledge that is based on the teaching of the microcosm, which regards the structure of man as a concrete analogy to the structure of the universe.

Understood in this practical sense, the law of analogy—which has also been called "the Great and Merciful Law of Analogy"—is more than a mere metaphysical construct. It is a guide to the experience of myself.

Death and birth exist on a cosmic scale. Therefore, according to the teaching of the microcosm, I may hope to discover an experience within the scale of my own existence that is analogous to death, but which yet does not reduce death to the level of my habitual egoistic reactions.

In some traditions, this search for an experience that is analogous to death is called "conscious dying." Numerous texts, which on the surface describe the efforts necessary at the last moment of life, may also be understood in a subtler way as a guide to the search for truth in the present moment of life. There is more than one meaning to the saying "A man may die at any moment."

Death is disappearance, coming to an end, the giving up of clinging.

But this too is only a thought. I see a person or even an animal die, and for a moment death is a concrete fact. In myself a separation between awareness and myself may take place; it is extraordinary. But it does not influence my life; it may affect my thought, but my thought has so little impact on my life.

My life goes on and I sink back again into the fears and lies that surround my attitude toward death. I cannot create an experience of death—but *can I not create this separation,* this moment of existing in the presence of myself? And is this separation of pure awareness from all that I ordinarily take to be myself an analogy to what has been called—in language which we no longer understand—the separation of "soul" and "body" in the moment of actual death at the end of life?

Here, perhaps, we have found an opening, a chink in the armor of the problem of death. But let us proceed cautiously. Nothing I can do from my own efforts can create that extraordinary state in which I stand for an instant face to face with the fact of death. Yet I can find an analogue to this experience. I can reproduce it in miniature and study the laws and the structure of my disappearance and the giving up of my clinging to my "self." *In a state of freedom from clinging to thought—where an important thought is allowed to move on, to disappear, forever beyond recall—one may have an experience analogous to that of death, which is the disappearance of my "self."*

It is of course for each individual to verify this claim for himself. And we may surmise that such verification is not so simple as it may sound, since in the literature of spiritual psychology the art of separation from thought is always associated with a discipline of bodily stillness and a certain precise balance between physical relaxation and tension. Moreover, such subtle bodily discipline is always immersed in the framework of a system of ideas, a relationship to a guide and a community, and to a certain attitude to life as a whole that must be studied and learned over a long period of time. Nevertheless, speaking for myself, I see in this direction the only ray of hope for remembering the awesomeness of death. And I am ready to listen to those teachers who say, either directly or in symbols, that the moment one sacrifices the egoistic relationship to a valued thought something instantly comes into existence in oneself that is like the birth of a new consciousness.

NOTES

1 Charles Singer, *A Short History of Medicine,* New York, Oxford University Press, 1928, p. 88.

2 Quoted in Ludwig Edelstein, *Ancient Medicine,* Baltimore, Johns Hopkins, 1962, pp. 451—52.

3 Rene Dubos, *Mirage of Health,* New York, Anchor Books, 1961, *p. 179.*

4 Evelyn Underhill, *Mysticism,* Cleveland, World Publishing Co., 1955, p 200.

5 Cited *in The Evolution of the Monastic Ideal* by Herbert B. Workman, Boston, Beacon Press, 1962, p. 143.

6 Paracelsus, *Selected Writings,* edited by Jolande Jacobi, Princeton, New Jersey, Princeton University Press, 1969, p. 81

CHAPTER THREE

The Science
of Living Being

We have become, in a painful, unwished-for way, nature itself. We have grown into everywhere, spreading like a new growth over the entire surface, touching and affecting every other kind of life, *incorporating* ourselves. . .We are now the dominant feature of our own environment. Humans, large terrestrial metazoans, fired by energy from microbial symbionts lodged in their cells, instructed by tapes of nucleic acid stretching back to the earliest live membranes, informed by neurons essentially the same as all the other neurons on earth, sharing structures with mastodons and lichens, living off the sun, are now in charge, running the place, for better or worse.

Or is it really this way? It could be, you know, just the other way around. Perhaps we are the invaded ones, the subjugated, used.

LEWIS THOMAS, *THE LIVES OF A CELL*

The world is a living creature endowed with a body which men can see and an intelligence which men cannot see.

ATTRIBUTED TO HERMES TRISMEGISTUS

Life Within Life

*A*lmost every great discovery of modern biology, every breakthrough to a new scale of size and time, reveals that life exists within life, and worlds exist within worlds. Every structure and process have shown themselves to be involved with the whole of life: from the digestion of food to the exchange of neuronal energies, to the patterns of insect communication, biological rhythms or bird migration. Whenever we have looked to a part for the sake of understanding the whole, we have eventually found that the part is a living component of the whole. In a universe without a visible center, biology presents a reality in which the existence of a center is everywhere implied.

The Crown of Evolution

If all of life is bound together, why is it, then, that the science of biology is so mute when we ask it about the meaning of *human* existence?

In the complex history of modern biology, only Darwin's theory of evolution has so shocked the mind as to raise serious questions about man's place in the universe. Darwin forced us to consider that we are animals, and that the designs of creation are played out on a much wider stage than was imagined. From the point of view of the theory of evolution, mankind is only one species among thousands which have their place within the field of organic life on earth. The fact that people took the theory of evolution as an enemy of religion only shows how rigidly they understood the idea of God. But the same may be said of scientists and others who now comfortably accept that the theory of evolution has disproved the Judeo-Christian understanding of man. As we shall see, this only shows how rigidly they have understood nature.

However, for most of us the theory of evolution has by now lost its shock value. Its place, in this respect, would seem to have been taken by the discoveries of molecular biology, the breaking of the genetic code and the promise of a new era of biological manipulation and genetic control. The previous generation was shocked by the idea that man was just another animal. One would think that we would be just as shocked to learn that essentially we are a computer program. *

Yet, the fact is that we have not been terribly disturbed by this information, which on the surface would seem to place us at a lower level of reality even than the animal. Apparently it is hard to accept that though we are by no means the pinnacle of God's creation, yet we are still part of a mystery.

*We leave out of account for now the fact that biologists are at the present moment in the process of discovering facts about the transfer of genetic information that bring the mystery and awesome complexity of purpose back into the intracellular processes themselves. Life inevitably is discovered to be composed of life. We always think we have finally isolated the mechanism (cell, chromosome, gene, DNA) only to find the mechanism is an organism.

But it has been rather easy to accept that we are automatons which operate according to logical rules we can understand with our logical minds. We would rather be computers with the *feeling* of power than living beings whose place we do not understand.

Scientists and others outdo each other portraying the promise and dangers of our newly found ability to manipulate life by rearranging the molecular structure of the gene. From one we learn that we will be able to breed humans with bigger, better or even extra bodily organs: a bigger brain, an extra thumb, or supernormal vision; another envisions the power to produce kinder, more loving human beings; still another recoils in horror at the thought of such manipulation and urges congressional action to forbid it. A noted political scientist soberly announces that "man is about to enter upon a new plane of existence," quoting with approval a leading medical researcher's description of mankind's new status in the cosmos: "The logical climax of evolution can be said to have occurred when, as is now imminent, a sentient species deliberately and directly assumes control of its own evolution."[1]

Nature's "Magic Wand"

In a presidential address to the American Association for the Advancement of Science, the eminent biologist James Bonner foresaw the day when "man will have the opportunity to literally remake himself in whatever image he chooses." And he gave an example of what for him will be an ultimate accomplishment:

> The brain today is about as big as we can handily carry about. If it were twice as large it would be quite a load. Even so, people of the future, who will depend even more than we do today upon full exploitation of their brain power, will doubtless want to have bigger brains. . . and this will be possible, because we will be able to leave them at home. With the development of sense organs for microwave communication there will be no reason why the individual sense organs cannot be made independent so that they can travel on their own, by microwave. The brain will stay at home, in a warm, comfy room, concentrating its efforts on thought, while the sense organs roam the world, seeing, talking, listening, playing—and continuously in

communication with the head office. We will enjoy a new freedom—freedom from carrying the head around.

Finally . . . we may very well give thought to the question of whether the gooey, sticky things of which we are made—nucleic acids, proteins, lipids and the like— are really the most suitable construction materials for such highly sophisticated, long-lived creatures as mankind will then be. People will say things like "Maybe a silicone backbone with four different markers would be better than deoxyribose and phosphodiester bonds. It would be less susceptible to cosmic rays and more resistant to attack by the strange new organisms which have just been found on the moons of Jupiter."[2]

To this glimpse of the future offered to us by molecular biology, I add one excerpt from among many such in Alvin Toffler's sensational, but well-researched report, *Future Shock:*

Dr. E. S. E. Hafez, an internationally respected biologist at Washington State University, has publicly suggested, on the basis of his own astonishing work on reproduction, that within a mere ten to fifteen years a woman will be able to buy a tiny frozen embryo, take it to her doctor, have it implanted in her uterus, carry it for nine months, and then give birth to it as though it had been conceived in her own body. . . Indeed, it will be possible at some point to do away with the female uterus altogether. Babies will be conceived, nurtured and raised to maturity outside the human body. . . Thus Dr. Hafez, in a sweep of his imagination, suggests that fertilized human eggs might be useful in the colonization of planets. "When you consider how much it costs in fuel to lift every pound off the launch pad," Dr. Hafez observes, "why send full-grown men and women aboard space ships? Instead, why not ship tiny embryos, in the care of a competent biologist. . . We miniaturize other spacecraft components. Why not the passengers?"[3]

On hearing these passages read to the class, one of my students excitedly suggested that here was a possible solution to the population crisis: miniaturize the human race! To my amazement, nobody laughed at this. Only I smiled uncomfortably.

I smiled uncomfortably because my mind was suddenly joggled into remembering a passage from P. D. Ouspensky's A *New Model of the Universe* in which he speculates about the remarkable organization of the social insects, suggesting that perhaps they were, so to say, an early experiment in consciousness, an experiment that failed. I had never been able to take that passage seriously. But now for a moment there loomed before me the picture of man, "the noblest of God's creations," as a colony of miniaturized beings whose central organ of intelligence was external to their bodies— bodies which were now as hardened as the chitinous exterior of the ants and bees, and whose mode of reproduction involved the fertilization of eggs outside the body.

It is impossible to become acquainted with their life without giving oneself up to emotional impressions of astonishment and bewilderment. Ants and bees alike both call for our admiration by the wonderful completeness of their organization, and at the same time repel and frighten us, and provoke a feeling of undefinable aversion by the invariably cold reasoning which dominates their life and by the absolute impossibility for an individual to escape from the wheel of life of the anthill or the beehive. We are terrified at the thought that we may resemble them.

Indeed what place do the communities of ants and bees occupy in the general scheme of things on our earth? How could they have come into being such as we observe them? All observations of their life and their organization inevitably lead us to one conclusion. The original organization of the "beehive" and the "anthill" in the remote past undoubtedly required reasoning and logical intelligence of great power, although at the same time the further existence of both the beehive and the anthill did not require any intelligence or reasoning at all.

How could this have happened?

It could only have happened in one way. If ants or bees, or both, of course at different periods, had been intelligent and evolving beings and then lost their intelligence and their ability to evolve, this would have happened only because their "intelligence" went against their "evolution," in other words, because in thinking that they were helping their evolution they managed somehow to arrest it.

· · ·

. . . They must have become convinced *that they knew* what was good and what was evil, and must have believed that *they themselves* could act according to their understanding. They renounced the idea of higher knowledge . . . and placed their faith in their own knowledge, their own powers and their own understanding of the aims and purposes of their existence. . . .

...

We must bear in mind that . . . every "experiment" of Nature, that is, every living being, every living organism, represents the expression of cosmic laws, a complex symbol or a complex hieroglyph. Having begun to alter their being, their life and their form, bees and ants, taken as individuals, severed their connection with the laws of Nature, ceased to express these laws individually and began to express them only collectively. And then Nature raised her magic wand, and they became small insects, incapable of doing Nature any harm.[4]

We have only to remember that Nature's "magic wand" need not be an external force, but an internal desire coupled with an extraordinary opinion about our own greatness and the inevitability of the steps we are about to take.

Upon what basis shall we choose between the molecular biologist's vision of modern man as approaching the level of divinity, and the suggestion that present-day humanity is potentially a race of cosmic insects?

Desire as Fate

Clearly, the modern ego has recovered quite rapidly from the shock provided by Darwin. In front of the discoveries of molecular biology, we feel burdened with great and fearful decisions involving the control not only of external nature, but of our very structure as living beings. We can do anything, be anything—all that we lack is the moral conviction to match our scientific power. Or, so say many of our scientists and social critics. A burden, but what a delicious burden! We are terrified of ourselves, but what a magnificent terror! No civilization in the history of the earth—not even, perhaps, in the history of the universe—has faced such a momentous decision. Truly, we are extraordinarily significant beings, we of the modern world.

I am by no means suggesting that as scientists we will not be able to do the things which some of us speak about. In the past, a foolish piety has gravely maintained that man would never fly, or synthesize protoplasm or accomplish some other marvel of manipulation. As pious critics of science, we have almost always been wrong. Nor do I mean especially to join the chorus of those among us who are terrified that we will use these discoveries immorally." No, the real terror is that we will continue to misunderstand the meaning of these discoveries as an index of our place in the cosmic scheme; and that, with egos inflated, by fervently pursuing this sort of science we will merely become ever more willing and ardent accomplices in the process of the physical and psychological "miniaturization" of man on earth.

The terror, in short, lies not so much in how we will use science. It is far more frightening to consider that the science we pursue is a result of our being at the mercy of forces we do not recognize, carrying us to a destiny that is far below our inherent capacities. Had we even in some small measure retained the esoteric idea of a multileveled, living universe, it might have occurred to us that the desires which our science serves are themselves a part of nature. As part of nature, these desires are related to other forces as effects are related to causes. These desires—the love of psychological security, the fear of bodily discomfort, the craving for individual recognition and for the feeling of power—at what level of nature do *they* exist? What sort of energies do *they* manifest? And are there other, more active energies within us which can manifest as quite a different sort of desire or striving, and which connect humanity to a more universal causal agency (a "higher purpose" in religious language) than that which now governs our fate?

Once again, we see the price we pay for our metaphysics. Without the idea of levels of consciousness in the cosmos, we confront a deterministic universe unwilling to apply that determinism to ourselves here and now. Without the idea of microcosmic man, we live as though our thoughts and desires are outside of nature. No matter what we claim theoretically about the universality of law, we contradict our own idea of universal law in the way we sense ourselves as "free" from moment to moment. Consequently, we never suspect that both our feelings of power and our moral tensions may be precisely the sort of force which connects us to the process by which man on earth could be rendered "harmless."

Is Nature a Pragmatist?

It is with neither moral outrage nor aesthetic horror that I view a future man reproduced outside the womb, with silicone spine and artificial organs. But I am astonished that we have become so stupid as to think we know so much about the real functions of our bodily organs. Is the heart, for example, only a mechanical pump? If so, then we must all yearn for the day when technology allows us to substitute a plastic heart with eternally replaceable parts for the heart we now have. Or do the tissues of a living human being contain substances sensitive to forces and purposes of which we in our present strange state of consciousness, are unaware? Surely we are at liberty to ask these questions now. We need no longer be intimidated by critics who take lightly the ancient understanding of the human body and its organs. For as soon as we introduce into our thinking the idea of a conscious universe and the human being in its image, we have no choice but to revalue every idea we have about the human body.

Frankly, what frightens me is not only that the acquisition of a plastic heart (I use this only as an example) would deprive us of possibilities inherent in the human organism. I am much more afraid that I would not even notice the loss, that my plastic heart and silicone spine would make me feel like a new man—so out of touch am I with the body and its qualities of energy. Could it be that we have already lost our possibilities, and that such changes in our structure will be the *result* of this loss of self-sensitivity, rather than its cause?

Because we are pragmatists with a two-dimensional metaphysics, we have actually taken nature and the human body to be pragmatists as well. If we observe that the heart pumps blood, we are inclined to think that must be all it does. If the eye is only a mechanism for the refraction of light waves, then we might agree with the nineteenth-century physicist Helmholtz who judged it to be so imperfect an instrument that, he said, had he purchased it from an optician, "he would have thought himself fully justified in returning it."[5]

But then, one asks, what more could the eye or the heart be than what our biologists tell us? I think the proper response to that question is not to cite some ancient idea (such as, for example, that the eye radiates a certain psychic energy) which, given our present understanding, would only lure us into fantasy or scorn about the ancient world's approach to nature. The task for us is to recognize that our understanding of nature may be a function of our level of consciousness, and that a new kind of biology is possible only on the basis of a change of consciousness. Let us proceed to think further along these lines.

Homocentrism

The great value of Darwinism, it seems to me, was that it jolted the modern world into questioning various sentimental beliefs about nature and our place in it. In this, Darwin's influence closely parallels that of Galileo. Just as the first modern astronomers and physicists destroyed a naive geocentrism, so Darwin and his successors overwhelmingly displaced what may be called *homocentrism,* the belief that nature exists for the sake of man.

As has been suggested, however, the geocentrism which modern science destroyed was the mere dregs of an enormously powerful universal idea that bears no resemblance whatever to the superstitious astronomical theory which we have taken it to be. I have tried to indicate that Galileo, for all his genius and integrity, was an instrument for the introduction into our lives of an esoteric idea—heliocentrism—which our modern world could not digest and which has been a primary factor in the spread of Western culture's metaphysical paranoia. Could it be that we have been just as hasty in our understanding of homocentrism, and with results just as destructive for a proper grasp of ourselves as living beings? Does Darwin, precisely because of his clarity and brilliance, stand with Galileo as a principal contributor to the modern world's misunderstanding of reality? Is the modern theory of evolution also only a distorted sacred idea?

Passivity Toward Nature

I had the good fortune of living most of my childhood on the border between a big city and a wooded area that was almost like an untamed forest.

Once off the trails one was surrounded by all manner of trees, ferns, rocks and flowers, as well as animals and birds in profusion. It was quite easy to lose oneself there, which I did more than once, wandering for hours until I found my way back. It was a deep and quiet place, and I needed no imagination to feel myself in the midst of nature.

I used to go there often, especially during my early teen-age years when I was keenly aware that I did not know why I existed. I would sometimes take a book, which I rarely read, and when I was in an unhappy mood I would simply tear into the woods, walking or running until I was tired, and then lie on the ground or curl up inside the hollow of some sun-baked rock until my emotions played themselves out. On quieter days I would sit by the river or under a tree and watch whatever form of life appeared— insect, squirrel or simply clear-running water—for hours and hours.

At that time I was also very taken by biology. I brought into the house every plant, insect and animal that my family would tolerate, graduating, finally, and to their consternation, to the breeding of innumerable tiny fruit flies in order to study the laws of heredity. But my family showed their approval of my interests by presenting me, at the age of sixteen, with a professional high-powered microscope. In the summer of that year I was one of a handful of young teenagers invited for intensive biological studies at a famous laboratory on the coast of Maine. There I was drilled in the principles and practices of scientific biological research.

All of this may suggest the picture of a young man searching for the meaning of life by communing with nature and studying science. But the truth is very different. I can say with certainty that in all those years nature *taught me nothing*. It delighted me, soothed me, relaxed me, and its beauty occasionally stimulated wonder in me. But, looking back on it and remembering many occasions since then when I have been in the midst of nature, I see that I did not know how to observe, I did not know how to be active toward nature. I never knew it was required. I always remained passive, relishing the feelings nature brought me, but sensing somehow, somewhere, without even admitting it to myself, that I had met nothing more real than my own feelings. Nature made me quiet, but soon after I became quiet I became bored and used nature merely as a backdrop for my imagination.

Living Ideas

But perhaps there was something else as well; there must have been. I was constantly reading about animals, insects and plants, and, as I have said, bringing things home to "study." And so I eagerly went off to be among the professional scientists for that summer. To my horror, I discovered—again without admitting it—that I preferred reading or hearing a lecture about the actions of the blood or the similarity of function between, say, a leaf and a lung, or between roots and intestines, to working in the laboratory, the marvelous, beautifully equipped laboratory where we were learning to become scientists.

At the same time, among my most vivid and profound impressions was the sight, through a microscope, of a blood cell or protozoan I had read about, or, in the midst of dissecting an animal coming upon a gland or structural configuration which had just been described in detail in the morning lecture.

The truth of the matter was not that I loved "nature," but that I loved ideas. And that I yearned to experience ideas *both* through my mind and through something else in myself that was not my mind.

I know now that when I learned about the structure of a leaf or the means of spore propagation in the mushroom, *I was taking in a real idea.*

Of course, I did not understand it that way then or for a long time thereafter. What were called ideas were the theories of the scientists—such as the theory of evolution, the theory of the gene, etc. These theories interested me very much, but not as much as I pretended to myself. What I really loved was the description and observation of natural phenomena. But these descriptions were given the names of *data, facts* by my teachers and peers. Only now do I see that in these descriptions I was coming into contact with *ideas* of a very different order than the theories of the scientists which purported to explain the data. Here we all were surrounded by the living reality of incarnate ideas, while being told to direct our minds to mere thoughts, ingenious though they may have been.

My teachers began to get impatient with me, and quite rightly for I was becoming careless in the lab. As for the theoretical work, my clever brain enabled me to hold my own there, but mainly out of a sense of competitiveness. Biology—my great love—was slipping away from me, so I felt. I was not really a scientist after all. Why? How did it happen? Was it because I was not a brilliant mathematician? Or was it because I could never feel that the mathematics we were being trained to use described the life which was right in front of us? I lost interest just at the point where, so I was told, the true scientist proves his mettle: the point where the quantification of observation is required. But I think the truth is that *the mathematics of modern biology is not the numerical expression of a living idea.* Only later did I hear that the ancient Pythagoreans used mathematics in this symbolic way, and only much, much later did I begin to grasp what that really meant.

Some critics of modern science argue that life cannot be measured and quantified. But to me that is a sentimental criticism and wide of the mark. It is not at all what I am speaking about here. I think that life, every living being, is an idea in time. If so, there must be a mathematics that expresses life. Simply because the metaphysical language of number has been abused by so many occultists and self-styled cabalists does not mean there is no way to express the precise and measured laws of conscious reality.

In any event, it is necessary for us to grapple with the ancient view of living beings as natural symbols: what might be called the laws of objective analogy, by which, in studying nature, I study the hidden structure of my own consciousness.

Creatures Are Symbols

I have long been intrigued by the idea that living creatures are symbols. To keep my attention on this idea, I must constantly resist the habitual tendency to believe symbols have a meaning that can be given solely in concepts. I am a child of the times. I must constantly struggle with the belief that knowledge comes into me through thought alone. Surrounded by the order and power of nature, I did not know how to be actively related to it. Like almost everyone else, I tried to be scientific through the associations of thought, which by itself is a passive and mechanical function.

I never understood why all the arguments among philosophers and scientists about purpose in nature seemed so hollow. What were they leaving out? What do we leave out of account in pondering this question?

The beginnings of an answer to this question came to me when I was studying the philosopher Spinoza. Here was a system more deterministic and mechanical even than modern scientific theories of the universe. Everything from God to man down to every creature in the cosmos was bound by iron law. Yet why did I sometimes emerge from reading Spinoza sensing *freedom* even in my body?

It was because from Spinoza I happened to receive mental impressions in a sequence and form that were organic and alive. The *Ethics* itself, though on the surface a geometrically ordered system of philosophy, is actually like an organism, moving, breathing, speaking, in its arrangement and order, its inner changes and conflicts resolved and dissolved, its thrusts and sudden halts, its gradualness, its cumulativeness. All this apart—though not entirely—from the content of what Spinoza said. In short, the *Ethics* acts to change my state of consciousness. Conversely, only in a somewhat different state of consciousness can I understandthe *Ethics*.

But this is precisely the sort of relationship I sometimes have with creatures, and which eventually alienated me from scientific biology. I did not know it then, but when I observed or listened to the description of a living being, it changed my state of consciousness—at least for a *moment*. But at that point something reciprocal was required of me, an effort of relationship toward myself, something similar to what Spinoza—and, even more, what sacred art and writing—requires of me. If we could understand that effort, I think we would understand what it means to be actively related to nature.

As scientists we do not move toward an effort at a new state of consciousness when we approach nature. Yet that is what is demanded of us if we are to understand the consciousness and purposes in natural reality. And it is surely this effort that distinguishes the relationship to nature which we see in traditional civilizations such as that of the American Indian or of the premodern Japanese. To understand intention and purpose outside myself, I have to have intention and purpose inside myself. Purpose cannot be experienced when I am absorbed in the automatic part of the mind, when my sense of myself is absorbed completely by the habitual associations of thought, no matter how "original" they may be.

Living beings have the power to change our state simply by our observation of them. Unfortunately, however, we fail to realize that this change of state lasts but a fraction of a second and immediately evokes a reaction in our ordinary psychology—what is called a change of *mood*. Walking through the woods as an adolescent, I settled always for a change of mood. Many of us are like that, and this is the source of all our sentimentality about nature, our naiveté, our subjectivity. A change of mood is not a change of consciousness. It is only another emotion, another feeling, more or less pleasant, more or less distracting, more or less in support of what I want to feel about myself. Moods bring with them certain perceptions and associations which are dreamlike in their subjectivity and unconnectedness to external reality. No wonder most of our nature poets, who are only expressing a mood, make connections between ideas and images that are pleasing or shocking, but rarely *true*. Let us call this subjective approach to nature *poetism*.

Scientism and poetism are therefore two sides of one coin. Sensing the subjective self-indulgence of poetism, scientism flees to ordinary, logical thinking. From the isolated emotions, it escapes into the isolated intellect. This was what my biology teachers demanded of me and it is what we demand of our young people when we educate them. Because we do not know how to do it or why, we offer them no way to return to themselves while they are turning to nature. We thus create a society of emotionalists and logicians eternally at odds with each other, "humanists" and "scientists" who vainly try to communicate about something, nature, which neither group approaches as a manifestation of consciousness. If there are "two cultures" in our society, this is the reason. The present conflict between the new religiosity and scientific technology is only the latest expression of our failure to grasp what it means to be actively related to nature.

The Loss of Symbolic Understanding

The view of man that arises out of modern biology is an inevitable result of our loss of the symbolic understanding of nature. The moment we forget that real symbols can be fully apprehended only in another state of consciousness, in that moment all the ancient teachings about man's place in nature begin to seem absurd. And we find ourselves assuming that before the nineteenth century everyone was slightly insane—or at best merely picturesque—about nature.

With little experience or appreciation of the state of being in which symbols are directly experienced, modern science has fallen back on another sort of analogy represented in the theory of evolution and the disciplines of comparative anatomy and physiology that are connected to it. It all stems from the fact that we do not sufficiently value the rare and momentary experiences of observing nature either directly or even through clear and comprehensive description. We therefore have no place in our thinking for these fleeting states which, it has been said, are the shadow of what is possible for the human understanding. And so, comparisons of bones, organs and physical functions lead us to the conclusion that man evolved out of the lower forms of life, when the only truly empirical conclusion is that our bones, organs and physical functions so evolved. We tacitly accept that ordinary thought and feeling are what distinguishes man. And surely man, understood thus, could very well have evolved out of the animal since such a being is little more than an animal himself. As one eminent biologist has put it, "Man is an animal plus something else,"—this "something else" being, in the eyes of this biologist, the power of scientific thought, a power which I am suggesting is in its contemporary manifestation largely a servant of fear.

We are bound to find it absurd when tradition tells us that every species of created being is the manifestation of an idea. We find it absurd because we do not discriminate between ideas and thoughts. Recognizing that our own thoughts are inadequate to reflect living being, we nevertheless fail to search for a higher quality of intelligence in ourselves. Instead, we pursue the search for a reality that is as mechanistic as the mind in which we habitually feel our sense of self. Yet every time we believe we have reached that level of reality (as many molecular biologists now believe) we discover that what we have found actually obeys the laws of purposeful life rather than the laws of automatism. Shifting from one biological scale to another yields various pragmatic results that make our lives easier. And perhaps it is that result which prevents us from staying longer with what presents itself on our own scale. *Instead of questioning the quality of our observing, we reach out for new things to observe.* To this end, we invent new instruments that reveal new worlds. But no instrument can give us symbolic understanding.

Tradition and Nature

Perhaps there were once peoples who learned directly from nature as from a sacred teaching, and perhaps somewhere even today there are still individuals who learn that way. If so, we would probably not recognize them and we would surely count as intolerable the danger or difficulty of their lives, which alone makes it possible for them to experience in their bodies the reflection of all the processes in objective nature.

Or perhaps such people never existed. For the fact, easily forgotten, is that all human beings that we know of live in nature on the basis of a system of ideas that is expressed in words, rituals, art and myth. It is only modern anthropology which leads us to believe that spiritual tradition can arise out of a people's relationship to nature as an effect arises out of a cause. We are so ready to believe that all human beings at all times were pragmatists like ourselves, and that every civilized form has the same *raison d' être* as do most of our recent forms: namely, physical safety and comfort or psychological pleasure.

Even the American Indian approached nature through the mediation of a revealed tradition. The Indian learns from nature to the extent that he learns from his religion. Nature teaches only him who is taught by God. That is to say, a spiritual discipline is needed in order for the conditions of a life in nature to evoke in man the state in which a symbolic, analogical understanding of the world is possible.

I have tried to offer the metaphysics behind this requirement in a previous chapter. The scale of time and intelligence in the objective universe is so vast that for man on this planet a spiritual teaching is necessary that brings down to our scale the whole scenario of cosmic reality. A great teaching in its entirety manifests the universe within the scale of human time.

If spiritual tradition makes it possible for human understanding to be fed by nature, it is therefore a great mistake to emulate a people's external relationship to nature without the inward able-ness to be open to the conditions created by their tradition. This is the mistake made by those who succumbed to the fantasy of the "noble savage," and also by those who seek to imitate either the American Indian or monks from other traditions who go off to live in caves and deserts. It is naive to select the part of a teaching one likes, while leaving the rest behind.

Without psychospiritual help, man in nature will generally sink to the level of the "animal plus something else." Such would be the fate of the "noble savage" and, in this sense, *modern science is an extension of the fantasy of the noble savage.* Proudly preparing to face nature directly without the superstitions and dogmas of distorted religious forms, scientific man needlessly cuts himself off from ever discovering the conditions of life created by undistorted tradition, conditions which alone offer him the possibility of assimilating into his being the total scale of natural energies whose transformation within his own organism can make him into the being which he now only imagines he is. This is the real meaning of the idea of homocentrism, and of the scriptural statements concerning man's lordship over God's creation. Man who is spoken of in the Bible and the Koran, who names the animals (through symbolic understanding) and who is designated as the vicegerent of God—*that man is microcosmic man.* He is not you and I, as we are.

The idea is that only the servant of God can be the ruler of nature. The "servant of God" is the being in whom the cosmic movement of energy outward and downward can take place without deviations and with sufficient intensity. The "ruler of nature" refers to the complementary ability of developed man to allow as well the movement upward and inward of these energies. Servant of God and ruler of nature thus characterize on a cosmic scale the two natures of man. Only such a human being has the power to take his real place and find his own will in the midst of all the universal forces acting upon and within him. To attach any other meaning to the phrase "ruler of nature"—to take it as referring solely to man's manipulation of animals or matter—seems to me a thoroughly egoistic and literal-minded error. Man has power over nature only to the extent that he has power—i.e., under-standing— over his own energies. But unperfected man, you and I, not only do not have this power, we are not even able to *see* these energies in move-ment within ourselves. For this seeing, a Path is said to be necessary.

According to the teachings of the Path, ordinary desire is a natural energy spending itself outward and downward through unperfected man. And scientific technology, to the extent that it is the production only of an intellect in the service of this desire, is little more than an instrument of this downward-directed natural energy. As we are, scientific technology is the badge of our slavery to nature. The irony is that we take it as a sign of unprecedented mastery and agonize over how to use it.

Man: A Symbol

Organic life on earth, while the manifestation of higher intelligence, points to that intelligence so mutely and indirectly that in itself it cannot teach most human beings or feed their being. Organic life is part of the wholeness of nature, but the universal intelligence it serves reaches into human consciousness *through the ideas of a sacred teaching* far more than through the rivers and the mountains. Without the help of real ideas, nature destroys man, or, rather, uses him only as an animal of a certain special kind, "an ani-mal plus something else." To love the song of a bird without loving even more the work of self-understanding, to love nature without loving true ideas, is to forget what it means to be man in the universal world, a being in whom two consciousnesses exist and who requires both extraordinary help and lifelong effort to build something in himself that can bring his two natures into relationship.

It may seem that our discussion has strayed far from the subject of biol-ogy. But I think it has been necessary, if only to suggest how far-reaching are the effects of our views about any fundamental aspect of reality. Thomas Aquinas once wrote that "a false opinion concerning the world will fatally engender a false opinion concerning God." This may not trouble us if by God we understand something external to ourselves. But if we set aside the customary associations of the word "God," and understand that this word also refers to our own psychological life's blood, then the issue is very different. It ceases being a question of merely opposing scientific views to the dogmatic utterances of a religion grown literal-minded or sentimental, and becomes instead a question of recognizing life-giving thought that both reflects the whole of reality and supports the struggle for self-knowledge.

In previous chapters I have maintained that as scientists we regard the universe in the way a literal-minded scholar regards a sacred book. The result is that instead of experiencing the unity of reality in our being, we live among concepts which preserve the fragmented world of appearances as it is structured by the egoistic personality. We live among intellectually resolved contradictions rather than among ideas that demand for their verification a deeper contact with our own inner life. The energies in the universe therefore pour through us as through a sieve. And that is what it means to live on the level of the animal. It is also what it means to be "mortal." The quality of our thinking has its influence upon the sort of experience we search for, which in turn eventually influences the way the forces of the universe move in us. This in its turn determines whether we live the extraordinary life of normal (microcosmic) man whose destiny, we are told, does not end with the death of the body, or whether we die as animals die.

Therefore, the main complaint we bring against contemporary biology is that it institutionalizes the absence of symbolic understanding and encourages us to approach the many-leveled reality of organic life with the cerebral intellect alone. It is not a question of wishing to be emotional about nature, but of reversing the state of affairs in which we are cut off from aspects of ourselves that resonate to the whole of creation. Symbolic understanding is an inner movement that corresponds to an outer reality and it is therefore a step on the way to man himself becoming a living symbol of the cosmos.

Examples of Symbolic Understanding

I wish to present some concrete examples of the symbolic understanding of nature, if only to emphasize again that I am not referring to poetry, "nature mysticism" or any of the other alternatives offered by people who complain that science is too cold. Several possibilities suggest themselves to me. Perhaps the most extraordinary example of a teaching that made explicit use of the symbolic understanding of nature was that of ancient Egypt.[6] Here in the sacred language (hieroglyphs) of the Egyptian tradition, the fundamental metaphysical principles (Neters) of the universe were discerned in surrounding nature and communicated through ideographic signs that could only be understood by one who had directly experienced the meaning they contained within himself. One instance of this must suffice:

The Sun and the Heart

The heart that gives your body life is a vessel of flesh, blood-distributor, *àb*, the eternal thirst whose rhythm governs your existence. It imposes this rhythm on your whole organism, as emanations from the sun, Râ, impinge on the "wanderers", the planets, that live in its orbit. Râ, heart of our solar system, like the heart that is the sun in our body, could never perform its life-giving part if its physical form was its sole reality, destructible because corruptible. But our texts always speak of the "indestructible Sun". When we adjure a dead man not to lose his heart, *àb* or hati, in the other world, we obviously don't have in mind the heart of flesh that is shut up in the tomb. The heart, like the sun, is the center of a world or system. It has, like the sun, two aspects, one visible and corporal, the other only perceptible by its effects. Aten, the solar disc, is but the physical body of the real star, centre of spheres of light, warmth and various powers. The heart of flesh, *àb*, is the body of that sun of life and fire which is the radiant centre of the soul, BA, whose lower aspect moves in the blood. Our true solar heart, our spiritual KA's centre of attraction, is the meeting place of everything in us that desires and accepts its impulses. It can steady and quicken the heart of flesh, which beats with its life: then the heart becomes a heart of fire, a centre of light, a source of life that has all power to subjugate the animal in us.[7]

Dissolution and Crystallization

In addition to the Egyptian tradition, one might turn for an example of the symbolic understanding of nature to the traditions of alchemy that are found in many cultures:

> In the world of forms Nature's 'mode of operation' consists of a continuous rhythm of 'dissolutions' and 'coagulations', or of disintegrations and formations, so that the dissolution of any formal entity is but the preparation for a new conjunction between a *forma* and its *materia*. Nature acts like Penelope who, to rid herself of unworthy suitors, unwound at night the wedding garment which she had woven during the day.
>
> In this way too the alchemist works. Following the adage *solve et coagula*, he dissolves the imperfect coagulations of the soul, reduces the latter to its *materia*, and crystallizes it anew in nobler form. But he can accomplish this work only in unison with Nature, by means of a natural vibration of the soul which awakes during the course of the work and links the human and cosmic domains. Then of her own accord Nature comes to the aid of art, according to the alchemical adage: 'The progress of the work pleases Nature greatly' *(opeis processio multum naturae placet)*. [8]

Nucleus and Radiation

Yet another example of the symbolic understanding of nature might be brought from India. Here is a brief selection from the writings of a contemporary master of the ancient system of Samkhya:

> If you consciously hold within yourself three quarters of your power and use only one quarter to respond to any communication coming from others, you can stop the automatic, immediate and thoughtless movement outwards, which leaves you with a feeling of emptiness, of having been consumed by life. This stopping of the movement outwards is not self-defense, but rather an effort to have the response come from within, from the deepest part of one's being. This process reverses the natural movement of *prakriti* (Great Nature) and brings back energy to its seed form. Let this become your way of communicating with others.
>
> ...

The law of life is the same. As the physical cells build the body, the germ cells are concentrated within and retain their energy for a later creation. We imagine that we create by projecting outwards, whereas real creation takes place through suction and absorption. When this power of absorption becomes natural, you discover that creation, radiation, communication and all similar processes come to you spontaneously.[9]

The Crucial Role of Biology

Many more examples could be cited from other traditions, the most obvious being that of the American Indian. But I think that quoting excerpts in this way is of limited use. Probably, what we have just read may better be understood as expressions or results of the symbolic understanding of nature. Results, that is, of a process in which impressions coming in from the external world are received within the organism in an order and by parts which correspond to the thing perceived. Symbolic understanding is, therefore, a process which can only take place to the extent that the knower is himself a moving symbol, a microcosm.

The cerebral intellect does not operate by this order of natural processes; it is logical, comparative and literal. The concepts which it produces support its useful function of computerizing impressions received in the more vital and organic parts of the human organism. Therefore, when symbolic understanding begins to take place within us, much depends on whether the cerebral intellect has at its disposal categories which do not contradict what the organism is mutely experiencing. Even more important is the mysterious quality of a man's willingness to suffer the temporary "derangement" of the logical mind when confronted with the "vital logic" of natural reality and the corresponding perceptions proceeding in him during those rare moments of symbolic understanding.

Now, if we put matters this way, we see that we are speaking of something upon which depends nothing less than the individual and collective future of man. A man's general attitude toward organic life reflects his general attitude toward that which surpasses his ordinary thinking. The science of biology therefore has an extremely crucial role to play in the evolution of human consciousness. A *wrong or shallow attitude toward living beings will infect a man's attitude toward those ideas which alone have the power to support the struggle for consciousness.* Just as true metaphysical ideas are necessary if man's being is to be fed by nature, so—to some extent—is the converse also true. To understand that nature is *like* a teaching is to be better prepared for the ideas and psychological forms of an actual sacred teaching. To put the point in dramatic terms: A superficial science of biology leads us deeper into an existence that is meaningless, violent and permeated with self-deception.

NOTES

1 John Heller, quoted in Victor C. Ferkiss, *Technological Man,* New York, New American Library, 1970, p. 99.

2 James Bonner in *Dialogue on Science,* ed. by William R. Cozart, Indianapolis, Bobbs-Merrill, 1967, pp. 80—81.

3 Alvin Toffler, *Future Shock,* New York, Random House, 1970, pp. 177—78.

4 P. D. Ouspensky, A *New Model of the Universe,* New York, Alfred A. Knopf, 1961, pp. 55—57.

5 Quoted in Garrett Hardin, *Nature and Man's Fate,* New York, New American Library, 1959, p. 59.

6 The recent studies of R. A. and Isha Schwaller de Lubicz have opened our eyes to this incredible tradition. See Isha Schwaller de Lubicz, *Her-Bak* (Vols. I and II), and R. A. Schwaller de Lubicz, *Le Miracle Egyptien,* Paris, Flammarion, 1963; *Le Temple de* l'Homme, Paris, Caractères, 1958; and Le *Roi de la Theocratie Pharaonique,* Paris, Flammarion, 1961.

7 Isha Schwaller de Lubicz, *Her-Bak: Egyptian Initiate,* London, Hodder and Stoughton, 1967, pp. 93-94.

8 Titus Burckhardt, *Alchemy,* New York, Penguin Books, Inc., 1972, *P.* 123.

9 Lizelle Reymond, *To Live Within,* New York, Penguin Books, Inc., 1973.

CHAPTER FOUR

Physics

The Influence of Modern Physics

A lthough the concepts of physics still exert a powerful influence on modern thought, that influence is now much less direct than it once was. Generally speaking, people no longer feel that if they really wish to get a picture of the universe they must master advanced mathematical theory.

The situation used to be very different. How many thousands of us grew up believing that the great Albert Einstein had with one simple stroke unveiled the nature of reality? Even as children, we would stare at the formula $E=MC^2$ as though we were looking into the very face of the Creator. And we felt sure that the physicists who understood Einstein's theory of relativity were separated from us not only by their intellectual powers but by their closeness to universal truth.

Of all the sciences, it was physics that told us we must think in new ways about reality. Darwinian biology and Freudian psychology may have upset our opinions about our place in nature. But physics alone seemed to challenge one's ability to *understand* the universe. Whether or not one agreed with a Darwin or a Freud, it was not so difficult to grasp the sense of their theories. The opposite was true with an Einstein or Planck or Heisenberg.

This is an extremely important consideration. *Physics defined for our contemporary mind what it means to have a new understanding.* We were told that the physicists were dealing with the basic laws of the universe. And we were told that our ordinary thinking was inadequate to grasp the nature of their efforts. In this way, we were given a model of what it means to think in a new way, and of what it means to see things from the perspective of a different scale of reality. Up until quite recently, to think in a new way about space and time meant to struggle to grasp the theory of relativity. And to think in a new way about causality and freedom required that one try to understand quantum physics or the rudiments of statistical mechanics.

The theory of relativity spoke of the dependence of the flow of time upon the frame of reference of the observer. It told us that there was no such thing as absolute time, independent of the act of measurement and of the physical factors involved in taking measurements of time. We learned, for instance, that a clock measuring time at a certain rate in one physical frame of reference would measure out time faster or slower when seen from another frame of reference moving past it. But since all things in the world, from galaxies to molecules, exhibited change or motion in one form or another, since all things, in their way, were "clocks," with "time" blended into their very essence, did that mean that the world we see before our eyes has no real structure apart from an observer taking measurements of it?

When the theory of relativity first brought us to such questions, most of us experienced a shock. And we may have responded to that shock by pondering anew what is actually the very ancient idea of *universal relativity*. According to this idea, all things in the universe exist only in relationship to a mind which perceives them or a purposive consciousness which creates them. On a personal level, this idea has through the ages led men and women to question the reality of everything they suffer for and reach after in the round of everyday life, and in the meshes of the everyday belief in time.

Yet such questions were frowned upon by most physicists, who quite understandably wished to keep science separated from the labyrinth of metaphysical speculation. "This is not psychological time, but physical time," some book might tell us. And turning away from a certain whisper in our minds, we found ourselves merely straining to follow the logic of verbal distinctions and mathematical equations. In short, the idea of the relativity of time, which often leads people to become serious about life and death, now led us only to question the accuracy of Newtonian mechanics.

The Question of a New Mind

I felt cheated without knowing why. Even when I grasped the mathematics of relativity physics and followed the texts some aching, unformed question about time remained absolutely untouched. Even though the books and professors were saying that Einstein cut through the riddle that had perplexed the ancients.

I wish I could say that by myself I plunged into the sacred texts of Asia and the Near East, where, as I much later realized, the mysteries of cosmic relativity are related symbolically to the inner life of man. I would like to report that I held fast to my sense of the incompleteness of modern physics. But I did no such thing.

So moved was I by even the merest brush with a universe beyond man's ordinary comprehension that I too fell in love with Albert Einstein and the theory of relativity. Along with almost everyone else a quarter of a century ago I thought of Einstein not simply as a great scientist, but also as a kind of spiritual prophet. Apart from his admirable qualities as a person, Einstein had made the structure of the universe an unknown once again. I mean that something in his theories reverberated with an unknown place in myself. Science was my "new religion." And in this new religion Einstein had the role of Moses, the great lawgiver.

It is easy for us now to criticize science for contributing so heavily to the modern sense of alienation in the cosmos. It is easy, and I think right, to call attention to the factors of human suggestibility which led so many of us to overestimate the kind of knowledge which modern physics brought. But, speaking for myself, I remember turning to the poets, artists and philosophers then in fashion without once sensing the demand that modern physics placed before us: *the demand to think in new categories* about the universe. One came away from the poets and novelists with heightened emotions about the human situation or about nature. But one never felt from them the demand to transform the mind in order to understand the laws of reality. I think that only physics had the power to evoke this need in people, though not even the physicists themselves appreciated the importance of precisely this aspect of their work.

Only physics *demanded* that I think in a new way about time and space, causality and freedom. Modern art could never reach into us in a way that made us feel a life-or-death command to understand time, for example, or energy. The horror of the atomic bomb, the death of thousands, millions, the possible destruction of life on earth, including my life on earth—all that was intimately bound up with the idea that there existed another scale of energy in the universe, as well as entities that obeyed different laws than the familiar things in the world we saw around us. It was therefore physics—not the arts, not religion, not philosophy— that *forced* us to question our understanding of reality. It did this by creating in us a connection, however transitory, between our fears and desires, and our concepts of the universe.

Of course, physics never saw its mission in this way, and it soon veered off from this quality it alone possessed. But the fact remains that for a time in our modern world only physics had the power to bring us closer to the search for a new structure of mind, a new consciousness, based on confrontation with the fact that I do not know what I am in this universe of immense pattern and incomprehensible force.

Levels of Ideas

We need to recognize that investigations into time, space, causality and energy can never be limited to the so-called physical world alone. Such categories apply to the whole of reality and therefore involve man in all aspects of his being. When Einstein's theories began to make people feel differently about their lives, many scientists rushed into print asserting that the theory of relativity was concerned solely with the measurement of external, experimental data. Often they heaped ridicule on people who used the theory of relativity as a starting point for self-examination or metaphysical reflection.

Nevertheless, it is impossible to hear serious, new ideas about time without coming inward, even if unwillingly and only momentarily. And here we touch on one of the strangest destinies of any group of modern thinkers. Every day physicists deal with concepts which in certain respects resemble esoteric ideas. Intellectually, these ideas throw into a new and larger context many of the assumptions about the structure of the world by means of which human beings conduct their ordinary lives. Intellectually, professionally, the physicist lives close to another reality, perhaps in its way a "higher" reality. But he has no hope of ever bringing these ideas into relationship with the whole of his life. The very thought of such a possibility would probably strike him as absurd.

Yet these ideas about space, time, causality and energy resemble esoteric ideas in that they put into question our everyday way of thinking about reality. To hear time spoken of as a fourth dimension, to be told that matter is a form of energy, or that causality does not exist at the nuclear level, is to feel the need not only for new thoughts, but for a new mind. It is to feel the need for an entirely new and free relationship to the thinking process itself— that same thinking process in which we feel our everyday sense of identity and which we trust in to guide us between the moments of crisis in our lives— the same thinking process by which we refine our values, set our goals, judge ourselves, direct our slender forces, estimate the thrust and meaning of our very existence on this planet.

I submit that Einstein's theory of relativity came into ken and affected the lives of people in this manner—for a short, but unforgettable period in modern intellectual history. Yet the result has not at all been to maintain this search for a new relationship to the thinking process—the search for what I am calling a new consciousness that is free to relate to thought without being swallowed by the chains of mental association that characterize the automatic psychic processes of the ordinary mind.

The history of modern physics thus contains a moment when two or three isolated ideas from another level entered into the stream of our culture, only to be completely engulfed by our habitual psychic automatism. Instead of discovering a new understanding of thought, we are now merely surrounded by new concepts, strange concepts, which are already becoming as familiar and "commonsensical" as those of the past century. Our psychological dependence upon associative thought remains.

It seems that it is always that way: We lack the ability to discriminate between new ideas and strange concepts. A new idea, one which comes from another level in ourselves, relativizes our relationship to concepts and explanations. That is, it causes us experientially to put in question the whole of our habitual thinking process. But a "strange concept," which masquerades as a new idea, but which is merely an unexpected shifting around of labels, words and images, serves instead as a magnet for the old quality of thought. We need to recognize this process by which a new idea degenerates into a new explanation.

Until now, I have been suggesting that modern man turned to ideas emanating from the disciplines of the Path without himself following these disciplines, thus turning the awakening force of great ideas into fuel for the engines of egoism. But the history of modern physics shows us that this process can also be mirrored in ourselves. In ourselves there appears from time to time a "great idea," a new idea that silences the associative circle of thought. What would it mean to bear this silence, which takes the form of the awareness of our own ignorance? Twenty-five hundred years ago Socrates walked the streets of Athens searching for individuals who could bear the sensation of "I do not know." But the record we have of his conversations with his pupils shows us that it is almost impossible for human beings to remain with the sensation of ignorance. Yet it was the message of Socrates that the new mind, the new consciousness which alone can understand reality, only appears when this inner price is paid in full.

We see, however, that we are afraid to distrust our thoughts. And so, imperceptibly and swiftly, explanations gather together and fill the emptiness created by the reception of a great idea. What is called a "new paradigm" or a "breakthrough" is then celebrated. But it may only be a new turn of the wheel of our bondage to the isolated intellect.

Strange Concepts

The paradoxes and shocks to "common sense" produced by the theory of relativity have by now more or less slid into the general fund of acceptable, if not widely understood, theories about the universe. Not so with particle physics—the branch of physics that deals with the dynamics and microstructure of the atomic nucleus. There is still a great deal of life and discovery, a great deal of interesting trouble for the logical intellect in this field. Unfortunately, we have all been so grim and tense about the paradoxes of subatomic physics—or, on the other hand, so whimsical—that here too we shall probably waste the help it has to offer in exposing the limitations of our understanding.

We have already made reference to the famous indeterminacy principle, the notion that at the level of the individual atom the laws of causality do not obtain. This is the major paradox offered by particle physics, and it has been with us almost as long as relativity theory. Like the idea of the relativity of time, the denial of universal causal law—that is to say, the assertion that there are some events in the universe that happen *without a cause*—is a great shock to the mind. Many physicists themselves, including Einstein, have not been able to accept the idea, although the theories in which it is embedded have proved pragmatically at least as useful as the theory of relativity. A famous debate about the subject was held in the 1920s, shortly after the idea was first proposed, between Einstein and the great Danish physicist Niels Bohr.

Einstein could not accept that an individual particle escapes from a radioactive nucleus *completely without cause*. He granted that at the atomic and subatomic level the only laws we could work with were statistical laws. But that, he said, was only because of our present ignorance of the causes of individual atomic events. Those who were with Bohr ("the Copenhagen School") argued, on the contrary, that at the atomic level individual causes simply did not exist. There things just happen. The only laws that exist are those of probability, which are quite as rigorous as any laws of classical physics. But *there are no laws* governing the movement of individual particles.

I cannot here summarize the theoretical problems and the experimental data which led physicists to this strange concept. Numerous clear nontechnical accounts[1] of the history of quantum mechanics describe the discovery of the electromagnetic field with its wave properties and the equally strong but contradictory evidence that light and electrical energy are transmitted in particle-like bundles called *quanta*. One learns how the German physicist Werner Heisenberg calculated that it was impossible to determine the simultaneous position and momentum of an elementary particle. This led to the conclusion—putting it very simply—that within certain limits the movement of an individual electron is governed solely by *chance*.

The interpretation of this indeterminacy principle is still a matter of debate among scientists. Many say, as Einstein did half a century ago, that it is only a question of our limited tools of measurement. Others maintain that this indeterminacy is an inherent property of nature itself on the subatomic level.

But the important fact for us is that modern science has introduced the element of chance into the universe, whether we interpret this as concerning the inescapable limitations of our knowledge, or whether we interpret it as applying to the things we know.

On the one hand, the whole enterprise of science is based on the assumption that nature operates according to rigorous causal laws. But on the other hand, the fundamental building blocks of this causally ordered universe do not themselves obey the laws of cause and effect.

Many ingenious intellectual maneuvers have been proposed to help us fight our way out of this contradiction. But the fact remains that after half a century modern physics presents us with a world view which explodes all the unified pictures of the universe that have been handed down to us from the past. What becomes of a universe harmoniously ordered by a divine will when chance rules at the heart of this universe? Some say that this indeterminacy exists at so small a level that it really does not interfere with the idea of causal order. But no one searching for a comprehensive picture of reality is willing to settle for a cosmos that is just a little bit contradictory, in which laws extend almost, but not quite, down to the core of things.

What to do? How to think? One might say that here is a problem that cannot really be solved by ordinary sense perception and logical thought. Here is a clear case where another quality of mind is required. I think there is great truth to that, and we shall go into it presently at some length. But before doing so, we need to ask if we have not been too rigid and simplistic in our ordinary understanding of how events happen right before our eyes, how things move, change and develop, on the ordinary level at which we live. Does not chance also exist in front of us every day, in everything that happens?

We have been speaking of a conscious, living universe. But everything that lives transforms disorder into order. Everything that dies moves from order into disorder. This movement between order and disorder, between unity and dispersion, between energy and manifestation—movement in both directions—is precisely the sense and meaning of a living universe, what the ancient Hindus called a "breathing cosmos."

And what does the ancient phrase "the will of God" really mean? Is it only a plan, static and unchangeable, which exists in the mind of some imagined supreme being? If so, it should have been called "the wish of God" or "the longing of God." Or does this phrase actually refer to a force as well as a plan?

We have taken the ancient, metaphysical idea of Creation and Return (movement toward and away from the Center) as far too automatic a process. A force must necessarily develop amid obstacles—or rather amid what acts at one level as an obstacle. Indeed, this idea that force requires resistance may be one meaning of the myth of Satan, who resisted the will of God, but whose very resistance was then utilized by the Creator for the fulfillment of the universal order on a larger scale. A similar meaning may be sensed in the numerous legends of divine personages who "fall," or who "make a mistake," as a result of which the cosmos can then come into being in all its vast patterns of movement and structure.

Do we see signs of this universal inherency in the events that take place right before our eyes? Does there come a point in the development of every process when *risk* enters, when *further development is not guaranteed?* If we put the question of chance in this way, we see that it involves the living relationship of processes to each other. The idea is that at a certain point in every process interaction with other processes takes place, an interaction which on one level appears as resistance, but which is actually necessary to the *full* development of the process. At the same time, such development is not guaranteed, but requires something "more," something "extra." The idea of man as a self-developing being seems to mean that, in all of nature, of him alone is it required and is it possible for that something "extra" to be called into action from himself, and not only from outside. I take this to be the objective basis of the idea that the development of man must involve the development of *will*.

How do these reflections connect to the problem of chance as it is brought by modern science? For one thing, they suggest that it is impossible to resolve the contradiction between causality and indeterminism solely on the basis of the sort of data which physics can provide. To think of chance as an opening rather than as a limitation requires a different view of process than we get from either modern physics or from the dogmatic cosmological systems of rationalistic religion and philosophy. Physics limits its perception of process to physical movement in space and time. Chemistry sees process in terms of change of attributes in the reaction patterns of known substances. Biology sees process in yet another way, and the sciences of human behavior see it in still another way. Thus, we very much suffer from one-sided points of view when thinking about the laws of change in the universe and in ourselves. Along with this, we suffer from the lack of a language which can refer to the whole of moving reality, rather than only to one or another aspect of it.

Modern physics has served us well in bringing us face to face with contradictions and fragmentations in our conception of the universe. But it does not serve us very well at all in offering what are only "strange concepts" that plaster these contradictions over, or which enable us to proceed "pragmatically" in the hope that someday, somewhere, these contradictions will be resolved merely in theory, rather than in the conscious experience of individuals. For myself, of course, I cannot say that the idea of chance as it comes across in physics brings me to the sense that life exists even on the atomic level. But I can say that it makes me question my old and stale understanding of what a living process really is. I see that I am unable to tell the difference between a process that is developing toward unity (toward the center) or toward dispersion and fragmentation. Unable to see this difference in the processes that take place before my eyes, I find, *we* find, that we disregard this difference altogether. Equally important, we cannot really tell the difference between processes that develop fully and processes that do not develop at all past a certain point. We have lost the sense of a whole process which comes into existence in time.

What are we seeing when we record the process by which the nucleus of an atom disintegrates or changes by throwing off its subnuclear components (the "particles" of nuclear physics)? To account for the observed behavior of these particles, physicists have reached for any number of notions which are in their way as "strange" as the notion of absolute chance, and which also tempt us to treat as external to ourselves aspects of reality that pervade the entire fabric of the universe, in eluding our own consciousness and inner functions. Concepts of subatomic "time-reversal," creation of particles out of nothing, parallel worlds of antimatter, and so forth, are invented and manipulated solely to account for the wild plethora of unexpected observations in the bevatrons and bubble chambers.

Universal Law and Self-observation

But these things—time, energy, causality—are part of *me*. Time —whatever it may be—pervades my body, my emotions, my experiences, everything that is in myself. And whatever energy is, it is obviously my life and movement, my power in all senses of the word. As for causality, can it be studied apart from observing in myself the forces that influence my own existence? Lack of such self-observation has led us to the strictly gratuitous conclusion that the psyche of man is separated from the cosmos around him. Shallow thinking about causality, how it operates, what it is, what it includes, is a factor that produced the modern myth of man's aloneness in the universe, a myth that serves at one stroke to make us feel both alienated and exceptional.

In the great spiritual disciplines of the world, the path of self-knowledge is precisely the study of time, energy and causality in *oneself*. For example, the ancient Hindu tradition of Samkhya speaks of the evolution and degeneration of energies in the cosmos, the expansion and contraction of time, the subtle or coarse qualities of the substances that enter into the physical and psychic functions of the organism—all of which need to be observed in oneself. The same is true of various schools of Buddhism and Sufism. But it has also been true of Judaism and Christianity.

Here, by the way, we touch upon a rather explosive idea. Even those of us who are willing to grant the symbolic nature of the Bible might balk at the notion that it speaks about such apparently scientific things as time, energy or causality. But that is only because we bring to these terms the associations driven into us by modern science. It therefore strikes us as merely bizarre to be told that, along with everything else, the Bible contains a teaching about the laws of the universe, the laws of Creation, destruction, the movement of time, the meaning of space and the structure of matter.

We read the Bible and in it we see religious teachings, moral tales, parables, histories and ethical imperatives. But perhaps the deepest meanings of scriptural language are only discovered through the observation of oneself. In the teachings of the East this aspect of scripture is easier to see—metaphysical language is constantly interspersed with the legends and spiritual imperatives.

No Hindu text, for example, is ever very far away from explicit mention of the three gunas, the three aspects of energy whose interplay creates the multitude of cosmic phenomena. Nor is there ever any doubt that these forces must be studied in oneself in the state of real self-observation called "meditation."*

---

* It is interesting to note that the "empiricist" philosophers, like David Hume, whose thought about knowledge and consciousness has exerted great influence in support of the scientific world view, base their conclusions on attempts at self-observation. Unfortunately, they never question their ability to observe themselves impartially, and in the modern world it has always been assumed—since the time of Descartes—that in order to observe oneself all that is required is for a person to "look within." No one ever imagines that self-observation may be a highly disciplined skill which requires long and subtle training and guided experience. In a strange way, then, much of modern scientific philosophy is based on fragmentary and unsustained self-observation. The later bad reputation of "introspection" (which is not necessarily the same thing as self-observation) results from the particular notion that all by himself and without guidance and training, a man can come to accurate and unmixed observations of his own thought and perception. In contrast to this, one could very well say that the heart of the psychological disciplines in the East and in the ancient Western world consists of training at self-study.

I do not wish to lay forth arguments for a specific line of biblical interpretation—one that would, for example, read the "justice" of Jahweh in
much the same way as the Hindus and Buddhists understand "karma," as a
law of action and reaction governing the entire reach of conscious and
mechanical energies in the universe; or that would read the "living water" of
the New Testament as similar to the spiritual energy that arises "from the
belly" in the Eastern traditions.[2]

The point I wish to raise is that we may be utterly mistaken in the way
we usually oppose the cosmology of the Bible to that of modern science.
The laws of the universe that spiritual traditions speak of may be laws that
can only be observed in a new state of consciousness. That is to say, they can
only be observed when a man is aware of the movements of energy *within
himself.*

This means that the laws of nature appear differently to us at different
stages of inner development. Biblical cosmology, therefore, is a call for a
developed attention to the universal forces both in man and in external
nature. It is a call for a new quality of observation, not necessarily a demand
for belief in statements that contradict sense perception.

I am suggesting that sacred language acts to evoke an echo in the human
psyche so that for a moment one glimpses a deeper law of reality—glimpses
it in feeling perhaps, or in some other inward movement that is as brief as it
is undefinable. Surely this is why we are told that there must be a certain seriousness or need, a kind of attentiveness that is the companion of need, when
receiving the formulations of a living teaching.

I think the foregoing is fundamental to understanding what it means
that sacred language is *symbolic.* Concerning the form in which knowledge
about the universe was transmitted in the ancient disciplines of the Path,
Ouspensky quotes his teacher, G. I. Gurdjieff, as follows:

> The symbols that were used to transmit ideas belonging to objective
> knowledge. . . not only transmitted the knowledge itself but showed
> also the way to it. The study of symbols, their construction and
> meaning, formed a very important part of the preparation for receiving
> objective knowledge and it was in itself a test because a literal or formal
> understanding of symbols at once made it impossible to receive any

further knowledge.

...

Among the formulas giving a summary of the content of many symbols there was one which has a particular significance, namely the formula *As above, so below,* from the "Emerald Tablets of Hermes Trismegistus." This formula stated that all the laws of the cosmos could be found in the atom or in any other phenomenon which exists as something completed according to certain laws. This same meaning was contained in the analogy drawn between the *microcosm—man,* and the *macrocosm*—the universe. The fundamental laws ... penetrate everything and should be studied simultaneously both in the world and in man. But in relation to himself man is a nearer and more accessible object of study and knowledge than the world of phenomena outside him. Therefore, in striving towards a knowledge of the universe, man should begin with the study of himself and with the realization of the fundamental laws within him.

From this point of view another formula, *Know thyself,* is full of particularly deep meaning and is one of the symbols leading to the knowledge of truth. The study of the world and the study of man will assist one another. In studying the world and its laws a man studies himself, and in studying himself he studies the world. In this sense every symbol teaches us something about ourselves.[3]

Seen in this way, a symbol is like a sound that causes certain psychological chords to vibrate without actually playing on the keys of the discursive intellect. The proper response to a symbol is therefore through an altogether new attention which can simultaneously move outward toward the object and inward toward the responses in myself.

Therefore, the world of nature can never become symbolic— that is, *meaningful*—to the ordinary unidirectional attention. If the universe is like a teaching, it also makes like demands upon the pupil.

The Contents of the Mind

Ever since the dawn of the scientific revolution there has existed in the West a thread of outcry against the reductionism of physics. As Theodore Roszak has pointed out in *Where the Wasteland Ends,* the Romantic poets—

Blake, Wordsworth, Goethe—rhapsodically inveighed against the universe of Isaac Newton: a universe of laws, forces and mathematical patterns. In the works of these and other poetic visionaries, ideas and formulae from a variety of spiritual traditions are brought to bear against the abstractions of mathematical physics. The poet demands a return inward to discover the riches within human subjectivity. It seems, in short, that even when science was rising to full tide the idea of microcosmic man was still alive.

Or was it? Does "movement inward" mean only an attention attracted passionately to the "higher" images and thoughts within oneself? If so, then Blake and the Romantic poets were indeed the prophets of a new vision of reality which many modern critics of science, as well as many troubled scientists themselves, proclaim them to be. But perhaps the movement inward that brings man in touch with fundamental reality is something quite different. I think it is. For in none of these visionaries are we helped to understand how to relate to the *contents of the psyche*—the fabulous and stunning patterns within the mind.

I wonder if what is really needed are images and conceptions that heighten our attraction to the contents of the mind, rather than ideas and a system of living that help us to understand the attraction itself and its psychophysical consequences. In any case, I am quite sure that it is the latter— freedom from the mind and emotional attractions—that is spoken of in the "psychological" writings of the great practical teachings of Asia and the Middle East. If so, if what is needed is the study of the forces that operate within the subjectivity of man, then it is quite wide of the mark to search for a rhapsodic imagery in place of the impersonal laws of physics.

Perhaps we are too ready to manufacture replacements for the conceptions of modern physics. How, instead, to look at the laws of physics themselves in a new way—so that they begin to "reverberate" with meaning? I do not think that an idea such as the microcosm gains its power by thrusting aside the laws of physics and drawing us toward the contents of the mind. That esoteric ideas have been misused in this way as a reaction to modern science I do not deny. We see it happening today, for example, in the way many followers of the new religions make use of Tibetan Buddhist texts and

methods—such as the four-sided mandala design—solely in order to intensify their interest in the contents of their own minds (when, in fact, the
whole aim of the Buddhist method is to free us from identifying with the
contents of the mind). Similar uses are being made of Hindu mantras (San-
skrit formulas), Sufi stories and the ancient legends of all cultures. But what
is gained merely by swinging from one attraction (toward things) to another
attraction (toward imagined things)?

I have already suggested that in modern times it was physics alone that
had the power to make us question our ability to understand the universe.
This it did not only by introducing extraordinary concepts of space and time,
but also by *making it impossible for us to turn to the contents of the mind for help in
understanding the laws of reality.*

Mathematics and a Foretaste of Consciousness

The word for this is *abstraction,* a mental process that is now justifiably
coming under fire by critics of modern science. But the effort of abstraction
is not necessarily an instrument of pragmatism, nor an end itself. In fact, it
may legitimately be seen as one kind of preparation for the struggle for consciousness.

The complaint one hears is that physics presents a reality for which no
physical models are adequate, that we cannot picture the entities and laws of
physical reality, that it is all a matter of difficult mathematical equations. I
feel the complaint is justified to the extent that physicists themselves try to
persuade us that mathematical abstraction is the sole means by which to
unlock the secrets of nature: But we need not give too much heed to how
the physicists value their work. As we have seen, they have always tended to
veer away from that aspect of their science which could bring us to fundamental questions about our place in the cosmic order.

The point is that mathematical physics is an echo of a central intellectual
power in man, a power which is spoken of in the traditions and with which
we have lost almost all contact in the modern age: the power to think without images, without words, without models based on the emotionally tinged
associations of everyday sense experience.

Physics

To turn to things that are presented to the senses and at the same time to see and—in the mind—to live through the laws which make up their being: that is a foretaste of another, deeply human work. I mean the work of being present in oneself to the movement of time and energy by which reality is created, manifested, maintained, destroyed and reborn within oneself according to laws which must necessarily exist in the cosmos.

Now one may ask: But what of the incredibly rich imagery of sacred scripture, allegory, myth, parable? Do not these communications evoke the emotions and stir the contents of the mind? Of course they do. The question is: in the context of what existential discipline? To be sure, one comes across these powerful images in every tradition and sees them at one level operating as lures for the pilgrim mind. But always and everywhere the listener is helped to see that they cannot be understood or made use of by themselves—that is, without the discipline. What discipline? Precisely the study of inner forces, fundamental universal forces. The contemporary love affair with ancient myth and legend thus often involves loving the means without pursuing the end for which they were established. The result is yet more attraction toward, and identification with, the contents of the mind.

To accuse modern physics of excluding mind and purpose from the cosmos seems to me only part of the issue. It is more accurate to say that the discipline of theoretical physics offers a foretaste of consciousness, *but does not understand or communicate the need for the development of consciousness.* Physics presents a world that is mechanical and lifeless even while demanding of the mind a more living attention. The mind comes alive yet what it sees is death. Without the help of ideas that awaken the search for self-knowledge and self-development, mathematics becomes a play of thought that can be turned to any use whatsoever. That may be why the spiritual symbolism of numbers was in the past reserved for those initiates in whom the need for consciousness had already begun to awaken. Almost all modern attempts to recover a lost Pythagoreanism neglect this point.

Let us return for a moment to the "strange concepts" of modern physics. It is now not the pragmatic usefulness of such concepts that interests us, but the liberty modern scientists feel in proposing them as the truth about objective reality. We see now that the reason lies in the disappearance from the modern world of the simultaneous study of oneself and the world of nature. And to repeat: By the study of oneself I do not mean the attraction to psychological images, associations, dreams, and so forth, but rather the direct observation in oneself of the universal laws of energy, time and causality.

Esoteric Ideas and Scientific Theory

The science of physics no longer carries the burden of questioning man's ability to understand the universe. That task is now being performed by the systems of Eastern religious thought which have captured the interest of so many Westerners, including a growing number of scientists themselves. A physicist in California, for example, is attempting to correlate and complement modern physics with the teachings of the Vedas of India and the mystical poetry of William Blake; another, in New York, sees parallels between quantum mechanics and the teachings of the Tibetan lama Chögyam Trungpa; yet another is attempting to exposit the principles of physics in the light of Hinduism as he has come to understand it through Transcendental Meditation.

Certainly this shift of thinking cannot be taken lightly, if only in regard to the enormous technological power of modern physical science. How to be sure that the ideas of Eastern religions are being understood rightly by individuals who have in their hands the conceptual tools and, indirectly, the machinery to alter the physical and biological environment of the human race? If a new range of ideas becomes the metaphysical guide of modern physics, what will be the effect on us should these ideas be accepted in the way we have been accustomed to accept and use great ideas? The ideas of matter and energy, for example, were originally formulated for Western civilization in the context of the Platonic, Christian and Islamic spiritual systems. If they have led to the use of the hydrogen bomb and to the modern illusions about psychological progress, then to what will these ideas lead in their Oriental formulations? And this is not even to mention other ideas which have the added attraction of relative unfamiliarity—such as the doctrines of cyclical time, the nonselfness of all entities and the cosmic force of "sound" vibration, to name only a few.

Obviously, we need to press this question. History shows us— and the traditions tell us—that good or evil for man appears in the space between what we know and what we are (that is, how we manifest what we know). For example, there is an old anecdote about the devil and a colleague of his who are walking along the street of a large city. The friend of the devil spies a man stopping and picking up something that has been thrown into the gutter. "Look at that man," says the friend, "how joyous he is all of a sudden. What was that he picked up off the street?" The devil replies: "A piece of the truth; he found a piece of the truth. But don't worry about it." "How can you be so complacent?" asked the friend. "Isn't that dangerous for you? Won't that hinder your work?" The devil smiles and walks on. "Not at all," he says. "When he gets home with that piece of the truth, I am going to help him organize it and apply it."

It is true that many scientists are taking up such Eastern practices as meditation and yoga. But is that alone enough to establish a harmonious relationship between sacred ideas and our everyday life in the physical world? Perhaps there is a wider gap than we imagine between the quiet attempt to observe the mind and the engagement with the material world that is demanded by the conditions of the modern life. And perhaps it is in that gap, in that space between quietude and activity, that the study of the microcosmic and macrocosmic laws of movement can take place—I mean the space between a more collected state of mind and the state one enters when struggling to deal with the problems of ordinary existence—scientific problems as well as social and personal problems.

Powerful metaphysical ideas may very well serve as a support toward a meditative state—in which state these ideas themselves are painstakingly verified through a subtle awareness of bodily sensation and the play of thought. But what becomes of these ideas in the transition to the situations of common life?

A slender, but telling hint is provided by the increasing interest that scientists and laymen alike are taking in the so-called "moment of insight," the "flash of intuition" that often accompanies influential scientific discoveries. A frequently cited example is the famous discovery of the form of the benzene molecule made by Kekulé in 1865. After struggling with this problem until he saw no way out, Kekulé one night dreamed of a snake eating its tail and awoke realizing the problem had been solved beneath the level of his ordinary thought. The discovery that organic chemical compounds take the form of rings was the basis of an entire branch of organic chemistry.

Many scientists have described to me their own experiences resembling that of Kekulé. I, too, have had such moments; I'm sure many people have. The question is: *What is of greater importance—the new explanation or the transformation of attention which has caused the new explanation to appear?* The attraction of the contents of the mind—an attraction in this case intensified by the pragmatic usefulness of the new explanation—is so great that we instantly pass from one state of attention to another without so much as noticing it, far less caring about it. In any event, we do so without knowing why or how or what it means that this transition from seeing to explaining has taken place.

This is a fragmentary illustration which I offer only to bring home the subtlety and importance of the problems we face in trying to make connections between esoteric ideas and scientific theories. It is essential to realize that science is an activity of ordinary existence, that scientific work takes place in the midst of the same sort of conditions and pressures that shape every other activity of our life. And so the disciplined study of the movement between inner states is basic to any possible connection between sacred teachings and scientific fact.

Verbal or logical resemblances, for example, between the cosmology of Buddhism—with its infinite universe pervaded by causal action—and scientific cosmological ideas count for little in this light. Who is there who is studying the process by which esoteric ideas are distorted? Surely such distortion does not begin with any explicit, easy-to-recognize alteration of verbal or logical content. Perhaps the distortion of truth begins with the transposition of an idea that can only be understood in a certain state of self-awareness to a context that is embedded in the conditions of ordinary life and the psychological state which these conditions invariably provoke in us.

This blind and unconscious transposition is egoistic pragmatism in a subtle and implicit form.

NOTES

1 For example, Andrade, Silva and Lochak, Quanta, New York, McGraw-Hill, 1969.

2 Cf. John 7:38. An excellent account of this idea is to be found in Hara, by Karlfried Dürckheim, London, Allen & Unwin, 1962.

3 P. D. Ouspensky, In Search of the Miraculous, New York, Harcourt, Brace and Company, 1949, p. 280.

CHAPTER FIVE

Psychotherapy
and the Sacred

The Era of Psychology

*M*odern psychiatry arose out of the vision that man must change himself and not depend for help upon an imaginary God. Over half a century ago, mainly through the insights of Freud and through the energies of those he influenced, the human psyche was wrested from the faltering hands of organized religion and was situated in the world of nature as a subject for scientific study. The cultural shock waves were enormous and long-lasting. But equal to them was the sense of hope that gradually took root throughout the Western world. To everyone, including those who offered countertheories to psychoanalysis, the main vision seemed indomitable: Science, which had brought undreamt-of power over external nature, could now turn to explaining and controlling the inner world of man.

The era of psychology was born. By the end of the Second World War many of the best minds of the new generation were magnetized by a belief in this new science of the psyche. Under the conviction that a way was now open to assuage the confusion and suffering of mankind, the study of the mind became a standard course of work in American universities. The ranks of psychiatry swelled, and its message was carried to the public through the changing forms of literature, art and educational theory. Against this juggernaut of new hope, organized religion was helpless. The concepts of human nature which had guided the Judeo-Christian tradition for two thousand years had now to be altered and corrected just as three hundred years earlier the Christian scheme of the cosmos retreated against the onslaught of the scientific revoltution.

Such, in quick strokes, is the background of a new question that is now arising concerning the hidden structure and distortions of man's inner life. A large and growing number of psychiatrists are now convinced that the Eastern religions offer an understanding of the mind far more complete than anything yet envisaged by Western science. At the same time, the leaders of the new religions themselves—the numerous gurus and spiritual teachers now in the West—are reformulating and adapting the traditional systems according to the language and atmosphere of modern psychology.

Taking his inspiration from elements of the Sufi tradition, psychologist Robert Ornstein writes: "We are now for the first time in a position to begin seriously dealing with a psychology which can speak of a 'transcendence of time as we know it.' . . .These traditional psychologies have been relegated to the 'esoteric' or the 'occult,' the realm of the mysterious—the word most often employed is 'mysticism.' . . .For Western students of psychology and science, it is time to begin a new synthesis, to 'translate' some of the concepts and ideas of traditional psychologies into modern psychological terms, to regain a balance lost. To do this, we must first extend the boundaries of inquiry of modern science, *extend our concept of what is possible for man.*"[1]

A book in itself would be needed to catalogue all the activity and theorizing now taking place among psychiatrists and psychologists attracted to Zen and Tibetan Buddhism, Sufism, Hinduism in its numerous forms and, lately, even the practices of early monastic and Eastern Christianity, as well as certain surviving remnants of the mystical Judaic tradition (Cabala and Hasidism). Added to this is the work of the humanistic and existentialist schools of psychology, pioneered by the researches of A. H. Maslow, which are now converging their energies on the mystical, or, as they call it, "transpersonal" dimension of psychology. Studies of states of consciousness, peak experiences, biofeedback, the psychophysiology of yoga, and "mind-expanding" drugs are more often than not set within the context of ideas and systems that hark back to the ancient integrative sciences of man. Finally, there is the acceleration of interest in the teachings of Carl Jung who from the very beginning moved away from the scientism of his mentor

Freud and toward the symbols and metaphysical concepts of the esoteric and occult.

Transformation and "Happiness"

Taking all of this together, it is no wonder that thousands of troubled men and women throughout America no longer know whether they need spiritual or psychological help. The line is blurred that divides the therapist from the spiritual master. As one observer, speaking only half facetiously, put it: "The shrinks are beginning to sound like gurus, and the gurus are beginning to sound like shrinks."

Yet I wonder if it is so simple a thing to pass off the distinction between the search for transformation and the desire for happiness. Is it only a matter of two different approaches to the same goal, two different conceptions of what is necessary for well-being, peace of mind or personal fulfillment? Or is it a question of two quite distinct directions that human life can take?

There is a fragment of an old Scottish fairy tale attributed to the pre-Christian Celts (the Druids, possibly) that tells of two brothers meeting on the side of a mountain. One is climbing and the other is descending. One is being led upward by a miraculous crane to which he is attached by a long golden thread. The other is led downward by a snarling black dog straining at an iron chain. They stop to speak about their journey and compare their difficulties. Each describes the same sorts of dangers and obstacles—precipices, huge sheer boulders, wild animals; and the same pleasures they have experienced—the wondrous vistas, the fragrant, subtle flowers. They agree to continue their journey together, but immediately the crane pulls the first brother upward and the dog drags the second downward. The first youth cuts the golden thread connecting him to the crane and seeks to guide himself by what he has heard from the other. But although all the obstacles are exactly in the places which the second brother has indicated, he finds each guarded by evil spirits, and without the crane to guide him he is constantly driven back and is himself eventually transformed into a spirit who must eternally stand guard inside a gaping crevasse.

The larger context of this tale is not known, but it may serve very well to open the question of the relationship between psychiatry and the sacred. Of all the numerous legends, fairy tales and myths that concern what are called "the two paths of life" (sometimes designated as "the path of the fall" and "the path of the return"), this particular fragment uniquely focuses on a neglected point about the obstacles in the struggle for awakening and the obstacles in the search for happiness. The tale is saying that however similar these obstacles might appear, in actuality they are very different and woe to him who fails to take them both into account. He will never move either toward "earthly happiness" or toward self-transformation.

This tale almost seems specifically designed to expose our present uncertainty about so-called "spiritual psychology." Consider the ideas, sourced in the practical core of the ancient traditions, that are now entering into the stream of modern psychological language: ideas about "states of consciousness," "enlightenment," "meditation," "freedom from the ego," "self-realization," to name only a few. Is it possible that each of these terms can be understood from two different angles of vision? For example, does one meditate in order to resolve the problems of life or to become conscious of the automatic movement of forces in oneself?

A similar question can be raised about the apparently non-psychological parts of these integrative teachings—the metaphysical and cosmological ideas which are often contradictory to the scientific world view and which are now being neglected or altered in order to fit in with contemporary associations of thought. Have we not seen by now that there are also two distinct human attitudes toward ideas—depending, to follow the imagery of our tale, on whether one is moving with or against "the pull of gravity" in life. We have been asking all along: What do ideas about the structure of the universe such as are contained in the Vedas, for example, or in our own Bible, demand of us? Why are they there? Why expressed in that way—surrounded by apparent inconsistencies and an often confounding imagery almost guaranteed to confuse or offend the "reasonable" side of our nature? And we have seen that on the whole our own attitude toward ideas is rooted in the search for explanations that serve the desire for satisfaction (intellectual or emotional). But there are ideas that are meant to be something other than explanations—ideas *that help us to discover the truth for ourselves as opposed to concepts that organize what has already been discovered either by ourselves or others.* Such ideas are what we have been calling "sacred" or "esoteric."

Our question here concerns psychiatry considered as a means to an end, as the removal of obstacles that stand in the way of happiness. (I choose the word "happiness" only for the sake of brevity; we could equally well speak of the goal of psychiatry as useful living, or the ability to stand on one's own feet or adjustment to society). These obstacles to happiness—our fears, unfulfilled desires, violent emotions, frustrations, maladaptive behavior—are the "sins" of our modern psychiatric "religion." But now we are asked to understand that there exist teachings about the universe and about man under whose guidance these "sins against happiness" may instead be embraced as material for the development of the force of consciousness.

At this point it would be helpful to pause briefly and recall our earlier discussion in Chapter One about the idea of the development of consciousness. There we introduced the age-old distinction between consciousness and the contents of consciousness and indicated that from the point of view of the Path, the great struggle is for an inner freedom that can simply watch and contain both the upward and the downward movements of energy within the psyche (the "upward" and "downward path" of our tale). The real enemy of self-development was identified as our automatic tendency to identify the whole of ourselves with one or another of these fundamental movements of psychic energy.

From this latter perspective, the main requirement for understanding the nature of consciousness is the *repeated personal effort* to be aware of whatever is taking place in the whole of ourselves at any given moment. All definitions or systematic explanations, no matter how profound, are secondary. Thus the formulations of both ancient masters and modern psychologists can be a diversion if they come to us in a way that does not support the immediate, choiceless awareness of the totality of ourselves in the present moment.

In traditional cultures special terms surround this quality of consciousness which connect it to the direct human participation in a higher, all-encompassing reality, "beyond the earth," as it is sometimes said. The existence of these special terms, such as *satori* (Zen Buddhism), *fana* (Islam), *pneuma* (Christianity) and many others, may serve for us as a sign that this quality of consciousness was always set apart from the normal, everyday goods of organized social and private life. And while the traditional teachings tell us that any human being may engage in the search for this quality of existence, it is ultimately recognized that only a very few will actually wish to do so. For, so we are told, it is a search that in the last analysis is undertaken solely for its own sake, *without recognizable psychological motivation.* Thus, what we have been calling the "esoteric" or inner Path embedded within every traditional culture is discoverable only by those yearning for something inexplicably beyond the duties and satisfactions of religious, moral and social life.

What we can recognize as psychiatric methods in traditional cultures must surely be understood in this light. Psychosis and neurosis were obviously known to the ancient world just as they are known in the few remaining traditional societies which still exist today in scattered pockets throughout the world. Surely, then, in a traditional culture the challenge of what we would call psychotherapy consisted in bringing a person back to a normal life without stamping out the nascent impulse toward transformation in the process of treatment. To do this, a practitioner would have had to recognize the difference in a man between thwarted normal psychological functioning and the unsatisfied yearning ("that comes from nowhere," as one Sufi teacher has described it) for the perfection of the soul. Certainly, that is one reason why traditionally the psychotic was treated by the priest. It is probably also why what we would call "neurosis" was handled within the once intact family structure, permeated as this structure was by the religious teachings of the culture.

It has been observed, rightly, I think, that modern psychiatry could only have assumed the place it now has after the breakdown of the family structure that dates back to the beginnings of recorded history. But the modern psychiatrist faces a tremendously difficult task as a surrogate parent, even beyond the problems that have been so thoroughly described under the psychoanalytic concept of transference. For there may be something far deeper, subtler and intensely human, something that echoes of another "cosmic dimension" hidden behind the difficulties and therapeutic opportunities of the classical psychoanalytic transference situation. We have given this hidden "something" a name: the desire for self-transformation. In the ancient family structure (as I am told it still exists, for example, among the Brahmin families of India) the problems of living a normal happy life are never separated from the sense of a higher dimension of human existence. What we might recognize as therapeutic counseling is given by family members or friends, but in such a way that a troubled individual may never confuse the two possible directions that his life can take. He is helped to see that the obstacles to happiness are not necessarily the obstacles to "spiritual realization," as it is called in such traditions. I think a great deal of what we take to be intolerable acceptance of restrictions, such as predetermined marriage partners or vocations, for example, or the fixed roles of men and women, may be connected to this spiritual factor in the make-up of the authentic traditional patterns of family life.

Can the modern psychiatrist duplicate this aspect of family influence? Almost certainly he cannot. For one thing, he himself probably did not grow up in such a family milieu; almost none of us in the modern world have. Therefore, the task he faces is even more demanding than most of us realize. He may recognize that religion has become a destructive influence in people's lives because the Path which the traditions offer has been covered over by ideas and doctrines which we have neither understood nor experienced. He may even judge that this same process of getting lost in undigested spiritual ideas and methods is taking place among many followers of the new religions. But at the same time, perhaps he sees that there can exist in people—be they "neurotic" or "normal"—this hidden desire for inner evolution. How can the therapist bring the patient to a tolerably normal life without crushing this other hidden impulse?

At the heart of the great traditions is the idea that the search for truth is undertaken for its own sake ultimately. These teachings in their entirety propose to show man the nature of this search and the laws behind it—laws which, as I have suggested, too often get lost in our enthusiasm for ideas and explanations that we have not deeply absorbed in the fire of living with all its sufferings and confusion. Psychotherapy, on the contrary, is surely a *means* to an end, what we have called happiness. Unlike the way offered by tradition, therapy is never an end in itself, never a way of life, but is motivated toward a goal which the therapist sees more clearly than his patient. The therapist may even experiment with invented methods to achieve this goal and often succeeds. But is it recognized that two kinds of success are possible in the process of therapy? On the one hand, the successful result may be a patient in whom the wish for evolution has been totally "disillusioned" and stamped out through the process of having had aroused and encouraged in himself the very quality of egoistic emotion which the traditions seek to break down and dissolve. But another kind of success may be possible in certain cases— a patient in whom the wish for evolution has been driven inside, who no longer longs for a response to this wish from the outside world, but who now has within him an even greater sensitivity and hunger for deeper contact with himself. To the outside observer such a person may seem to have developed a certain strength or "inner-directedness," but in actuality he is precisely the sort of person who may desperately need what the traditions seek to communicate. The efforts of contemporary teachers from the East to bring their message to such people in terms that are neither freighted with dead antiquity nor compromised by modern psychologisms constitute the real spiritual drama of the present age.

I suspect that many psychiatrists sense there can be these two different kinds of success in the process of psychotherapy. But the second class of patients probably leave the therapist before the treatment is far advanced, while the first class of patients stay and pay as long as they can. Therefore, this second type of patient is probably not consciously or officially recognized by the profession of psychiatry.

Ideas as Help

What is "help" for man? As we have said, during the first half of the twentieth century it was chiefly the Freudians who were offering the public a new answer to this question. But what is never said about Freudianism, or about the whole of modern psychology, is that the help it offers is directed solely to the automatic part of the human psyche. In psychoanalysis and its present-day derivatives the effort is chiefly to reawaken the intense, but nevertheless still automatic, emotional reactions which are often associated with childhood experiences. The automaticity, as well as the integrity, of childhood is understood as related to the most real part of a person.

Neither the Jungians nor the later Freudians ever completely abandoned this point of view about the importance of childhood reactions. Yet among the founders of modern psychology there existed practically no discrimination of levels with regard to the influences which produced these childhood reactions. To the modern psychologist it is simply assumed that the fundamental influences upon the child concern its physical safety, its recognition by others, its security and its bodily sensations. Parental love or its absence is understood entirely in these terms. The differences among the various schools of psychology depend in the main on which of these influences is understood as paramount in the formation of the individual's identity.

Other influences—upon the intellect, for example—were understood as secondary or derivative. It was never seriously considered that there might exist in the child a subtle emotional receptivity on the basis of which the ancient structures of traditional family life and education were built and maintained throughout the centuries. No attention was given to the communication of ideas through art or literature; nor to the possibility of essential learning experiences through the child's encounter with difficulties that demand he be more "present," that is, less automatic.

It would be wrong, of course, to lay too much blame on modern psychology for its estimation of the sort of influences that act upon us and govern our development. For in this it is only following in the wake of the practices of religion, art and education in the modern era. The ever-increasing tendency of Western religion to offer rationalistic explanations of its ancient teachings along with simplistic consolations concerning death and the problems of social and family life made it all but impossible for the influences of Christianity and Judaism to challenge the automaticity (or passivity) of human nature. Another word for this is "secularization"—a term which should, I believe, be understood in reference to the psychophysical structure of the human organism. Ideas and practices which were meant to penetrate behind the screen of man's automatic thoughts and motivations were so formalized or so covered over with rationalizations that they almost totally disappeared into the vortex of the ordinary influences of "civilized" life.

Similarly with education. The liberalizing of education in the modern era came from an intention to widen the scope of human knowledge and open the doors to every sphere of learning. Yet this movement has been so connected with the industrial and technological forces of modern life—that is to say, with the influences of pragmatism and materialism—that even the most powerful and all-encompassing teachings of mankind were presented to the automatic part of the mind, the part that merely remembers and recombines words and impressions like a computer. Gradually, the experience of learning became limited to the cerebral intellect alone and was valued solely for its results in terms of self-satisfaction or the "mastery" of nature in one form or another.

But what is admirable in the child is his integrity—the almost total unity of his innocent mind with his body. *That something of this organic unity should continue to exist as his mind is informed about life on this planet seems to me to be the basic aim and problem of education.*

As for art and literature, they have lost entirely their function of, first, awakening and supporting the wish for a new knowledge of unity and reality and, second, of transmitting subtle emotional impressions through allegory and symbol in an order more corresponding to the interconnected and causal world behind the appearances. Even the very possibility was forgotten that there can exist in ourselves a purer or less automatic emotionality that is actually an indispensable organ of knowing. Art was for "aesthetic pleasure" or persuasion, and then led in a different direction than knowledge.

Modern psychology has come to take its place as one of the numerous channels for influences that answer our desire for such goods as health, safety, satisfaction and the feeling of individual worth. As we have said, modern psychotherapy often works by supporting in the patient the very kinds of emotion that traditional spiritual teachings and methods seek to weaken. But from the point of view of what we have been calling the Path (the "upward path" of the legend), it is not simply the object of certain emotions or their painfulness that makes them a pathological factor in human life, but their automaticity and their power to mask the subtler and more active feelings that bring an individual in contact with reality. Thus, it is not a matter of what I desire (it may even be "God") or whether I suffer, but rather what in me is struggling to transcend desire as such.

Historically, sacred disciplines have offered themselves as a means for transcending the egoistic emotions for the sake of developing consciousness and moral force. As I understand it, these inner disciplines thus exist in the world as a completely unique sort of influence, unconnected to the help offered by psychotherapy, as well as by most modern forms of religion, art and education.

When seen from this point of view, it seems to me that none of the more recent attempts to broaden psychology have hit the mark. The numerous forms of "existential" psychology offer devastating criticism of the concept of the unconscious introduced by the depth psychologies. Yet neither they nor the depth psychologies entertain the possibility that there may indeed exist an unconscious intelligence in man, but one which is not "located," so to say, where it has been sought.

According to the teachings of the Path, this hidden intelligence comes into being and grows into an active force only through a specific form of voluntary struggle. And "sacred" ideas are intended not only to attract that hidden intelligence but therefore to support the struggle against attachment to all forms of psychic automatism, especially on the level of emotion and feeling. We are thus asked to understand that a "great idea" is not simply an idea about "great things" such as the soul or freedom or the cosmos.

The point is, rather, that in order for real ideas to be an influence against the rule of inner automatism, their expression must not only appeal to the nascent, hidden intelligence, but must also specifically interrupt the habitual, associative processes of thought and ordinary egoistic emotion. I see no other way of understanding why such ideas are rigorously set apart from the more familiar formulations about God and man that gradually become the comfortable heritage of particular periods of history under the name of religion and philosophy. In this way, one understands as well why such ideas are, part of the practical method of effecting palpable changes in the psychobiological organism of man, and why their influence in any particular era is only partly visible to many of us.

Is Love an Attribute of Reality?

But it is time to call things by their proper names and admit that we are speaking here about love—love both as a cosmic and as a human phenomenon. Surely one cannot speak for long about the idea of the microcosm without attempting to think about the meaning of love. A conscious universe, a universe that is more, not less, than man must contain love. The difficulty, the enormous difficulty, arises as we come to see how shallow, crude or sentimental is our understanding of love. Even people willing to entertain the thought that there is great intelligence and purpose in the cosmos balk at the possibility that love is an attribute of reality.

And well they might. Does not even the simplest observation reveal the "cruelty" or, in any event, the rigor and severity of nature? And is there anything more cloying than a metaphysician or nature poet who sees only "beauty" in the cycle of reciprocal killing and feeding that proceeds endlessly among all living things great and small? What love is there in that?

I remember very well the shock of relief I felt when I first read Freud on this subject long ago. It was in his book, *The Future of an Illusion*, which, along with *Civilization and Its Discontents,* so powerfully and cleanly defined the issue for our era. There he writes of religious ideas as being "born of the need to make tolerable the helplessness of man.

> It is true that nature does not ask us to restrain our instincts, she lets us do as we like; but she has her peculiarly effective mode of restricting us: She destroys us coldly, cruelly, callously, as it seems to us, and possibly just through what has caused our satisfaction. It was because of these very dangers with which nature threatens us that we united together and created culture, which amongst other things, is supposed to make our communal existence possible. Indeed, it is the principal task of culture, its real *raison d' être,* to defend us against nature.

One must confess that in many ways it already does this tolerably well, and clearly as time goes on it will be much more successful. But no one is under the illusion that nature has so far been vanquished; few dare to hope that she will ever be completely under man's subjection. There are the elements, which seem to mock at all human control; the earth, which quakes, is rent asunder, and buries man and all his works; the water, which in tumult floods and submerges all things; the storm, which drives all before it; there are the diseases, which we have only lately recognized as the attacks of other living creatures; and finally there is the painful riddle of death, for which no remedy at all has yet been found, nor probably ever will be. With these forces nature rises up before us, sublime, pitiless, inexorable; thus she brings again to mind our weakness and helplessness[2]. . .

As is well known, Freud considered religion to be a psychological gambit by which earlier cultures personalized the objective force of nature in order to feel more comfortable, or at least more emotionally intimate with great reality. The belief in God and a divine Providence was an illusion, a wish fulfillment. For Freud, the true fact about nature was that it was indifferent to man's needs.

In science, civilization had finally found its real champion in the conquest of nature. No longer was it necessary for humanity to postulate illusory gods in order to feel that there was something it could *do*. It would now be possible *actually* to master—though never completely—the forces that threaten human survival.

Psychoanalysis was put forth as the effort to perform more successfully the task that always defined the essence of civilization: to master the forces of nature in ourselves, though again never completely. Civilization only existed to the extent that man went against his own nature and delayed the gratification of what Freud called "instinct." It was man's relationship to the sex drive—conceived now very much as pitiless nature itself working in man—that determined his success or failure in surviving.

In psychoanalysis Freud saw that the task of science was to maximize the gratification of sexuality while maintaining the rational order of civilization as it was reflected in the individual.

Despite all the modifications and refinements that Freud and his followers made in his theories, the impact on society was quite elementary: Love was based on sexual pleasure and sexuality was blind nature's operation in the human animal. To an age uncertain about the meaning and value of art and religion, an age unable to take seriously the idea of higher knowledge, Freudian theory was irresistible. Had anyone answered Freud by saying that the existence of certain ideas and their expression in certain forms of art, philosophy and religion was itself the trace of another sort of love emanating from a stratum of reality that was hidden to modern man, he would have been ridiculed.

I was a young man when I first read Freud. Like almost everyone else, I had no means of distinguishing great ideas from concepts that were the product of ordinary minds like my own. Similarly with regard to art and religious thought. And, like many people, I suffered from a sort of mental and emotional malaise when pondering the basic questions of human existence. The teachings of the great masters mingled and disappeared among the arguments, fantasies and logical constructions that my own brain invented or that I borrowed from the writings of others. Therefore, the stunning clarity that Freud's theories brought. Immediately, I cut away all that malaise. Religion was an illusion. Man was an animal. Civilization was based on contradictions between the social and biological elements of human nature. It could never have occurred to me then that the contradictions in human nature which Freud and modern psychology perceived so emphatically could be understood in another way—in quite another way, as an expression of the complexity of the human organism and the fate of its lower parts, including ordinary reason, when they are severed from a relationship to higher levels of awareness.

What we may call psychological pragmatism—modern psychology—is the attempt to help an individual make peace with his contradictory nature without providing him real access to an awareness which alone has the power to bring order into the human chaos, and which alone is the manifestation of love toward and in man.

Whether it is Jung I read, Skinner or Maslow, it is always a matter of increasing pleasure, meaning or vitality within the contradictions of human nature. Neither the "peak experiences" of Maslow nor the "integration" spoken of by Carl Jung communicate the sense of man as a being between two levels of reality. It does not matter that one psychologist has rejected Freud's view of sexuality or that another speaks of transcendent experiences or that yet another brings in ideas from ancient spiritual and esoteric traditions. It is strictly a matter of help for man being understood by psychologists as the improvement of a horizontal cross section of man.

I wish to speak about this more personally because here, more than anywhere, concepts alone are of little help. I especially do not want to give the impression that I am criticizing psychology for leaving out the so-called "soul" of man.

A Glimpse of Objective Love

I can be quite brief. It happened not long ago, through a very unusual set of circumstances, that I was stranded without a cent in a foreign country. I knew absolutely no one and had no way of getting money or help until the following day. There was nothing to do but beg.

At first I treated it as a sort of lark and had no success even when I screwed up enough courage actually to speak. I did not look needy and I certainly communicated no urgency. People shied away, and I became curious as to why they turned away from me. At the same time I was beginning to experience the rare sense of myself that can appear when one is suddenly free from the habit of weighing courses of action. I was doing the only thing possible at the moment; there was no "hope," no possibility in anything else. There was a great deal of fear in my body, but no indecision in my mind. In short, I was gradually approaching a different state of awareness, although I would hardly have put it that way at the time.

Eventually, of course, I began to get hungry, quite hungry. My curiosity about people's behavior had become transformed into something much deeper and quicker. I began to feel in the very muscles of my body a need to get at them in the right way. Added to the problem was the fact that it was a holiday and I was now in a residential neighborhood. I had no choice but to knock on doors and ask for food.

Still no success. At each house it was a tremendous burden for me to actually ring the bell and speak. Very few people opened the door and of those who did, only one or two showed any courtesy. The rest shut the door with a hasty refusal. I couldn't believe the rudeness and callousness of people. My reactions moved back and forth from anger to laughter.

At this point I was really in need of something to eat, yet I still felt quite contained since I knew that tomorrow everything would be back to normal. My state was such that every impression was vivid and clear; I experienced rich sensations of myself walking, stopping. Sights, sounds and smells came alive. At the same time I felt my own helplessness and lack of intelligence—but mainly my profound inability to manifest toward people in any but the most habitual ways.

I was walking in a beautiful winding lane with the late afternoon light filtering down through the tops of huge old trees. I saw a woman with graying hair trimming the bushes in her garden. She was bending over right by the fence and I came almost within arm's reach of her. "Excuse me," I said. "Could you please give me something to eat?" There was a new note in my words. My voice came from a deeper place.

She stood up, surprised. I could not take my eyes from her face. She stepped back, tentatively, and I watched with extreme clarity as her features registered a certain now-familiar harshness coupled with a mechanical attempt to smile as she quickly looked me up and down. From the very first, I saw that my request was causing her to suffer. "Are you really hungry?" she finally asked with a kind of disbelief, still eying me all over. I answered that I would deeply appreciate anything she could offer me.

Now comes the point of this tale—and I wish I had the words to communicate more than a small fraction of what I saw. For quite a long time—perhaps thirty seconds—she struggled with herself. She seemed unable to move; now and again she would start to say something and stop abruptly.

What was she struggling with? Obviously, fear. Fear of I don't know what: not of me, but fear of moving against some deep habit. At the same time her eyes reflected a longing to be different, to be kind, to help. I am not stating this as a guess or an inference. In the state of mind I was in I *saw* it, more clearly than I see the paper in front of me now.

As I have said, she remained in that condition for quite some time. Finally, the words came out of her mouth brusquely, almost nastily: "I'm sorry. I don't have anything!" She snapped to going about her business, turning away from me. But I stayed awhile, watching her, before walking away. To my astonishment, I was gripped by an extraordinary feeling of love for this woman.

I have never, before or since, experienced love toward another living person in a way even remotely resembling what I felt as she was struggling with herself and succumbing to turning me down. I had always believed that objectivity or impartiality toward another human being meant that one was without feeling. The contrary is true. I would now say that without this glimpse of impartiality there is no such thing as love.

Love as an Attribute of Consciousness

This little incident, so important for my own understanding, carries many serious questions with it.

What state of consciousness is required for there to occur a true perception of another human being? For me it could not have happened without the force of exceptional circumstances that I could never have wished for in advance: genuine physical need; being thrown out of everything habitual; the absolute demand to act, together with the repeated impressions of my inability; freedom from worry; absence of real danger and much else, I am sure, that I cannot even guess entered in as contributing factors.

I must add that before this incident I had read as many books as the next person in which love was linked with knowledge. Now it seems incredible to me that people can write so glibly about these things. Various religious and psychological authors make it sound as if everyone has or can have a complete store of experiences where love and knowledge of another person are joined. Of course, I can only speak for myself, but if I should have even a small handful of such experiences in my lifetime I would count myself fortunate.

For a moment, I knew that woman. That knowledge was absolutely unrelated to any other occasion in my life when I thought I understood another person. Certainly, my experience as a clinical psychologist never prepared me for anything like it. There was no sense of *recognizing* the other person on the basis of psychological concepts, there was no feeling of what we call sympathy, there was not the slightest element of sexuality, there was no impression of my having accomplished something, no sense of a solution to a problem, not even an interest in the woman's personality or in getting to know her better. In that moment of love, I can honestly say that I did not even wish her to win her struggle and give me food.

Moreover, it stands to reason that although this experience seems so extraordinary to me, it would be absurd to regard it as anything but a fleeting glimpse of the purer emotions that are possible for human beings. After all is said and done, that is surely the one thing that needs to be remembered when critically evaluating ourselves as psychologists: How without noticing it in the slightest we accept the phenomenon of love on so small a scale. Simply to recognize that there may be such a thing as love on an entirely new scale is perhaps the bridge we need to approach the idea of cosmic love.

I do not think even the most "transcendentally" oriented new psychologists communicate this sense of cosmic scale in their reflections on love. We are so easily excited by one or two unusual experiences that, far from opening us to question the limitedness of our emotions, they almost always tempt us to believe we have purified our emotions. Whether we call them "peak experiences" or "breakthroughs" or "I-thou" encounters, our habit is to overestimate these experiences by believing that we have made contact with a higher level in ourselves. I sometimes wonder if, apart from everything else, this fact does not explain something of our overestimation of the sexual experience. Freud thought that sexual pleasure was the most intense experience that is possible for human beings and succumbed to constructing a whole theory of the psyche and civilization around this belief.

What Evokes Objective Love?

I ask myself: What was it in that woman that evoked this purer quality of love? Granted that a number of exceptional factors had produced in me an unusual breadth of awareness, still it was something in *her* which heightened it even further or, rather, which called into existence a quality of relationship that I had never experienced.

I say that it was her struggle with herself. She was *in between* in a way that we so rarely are or that we can so rarely tolerate. Was it that something flowed through me toward her because in her way she was struggling to be master of her habitual psychology?

Thus we are led back to the question: What is help for man? From where does it come? What must we contribute in order to be open to receive it?

Creation and Destruction

Many speculative philosophers have written about the idea, contained in many traditions, that in the universe there is a law of creation and a law of destruction, a law of love and a law of death. We find it clearly stated in Hinduism where these two forces are symbolized by the gods Brahma and Siva. In the Chinese system, Yang and Yin are sometimes understood in this way. In the Cabala, *Hesed* may be taken as the force of God's creative love and mercy, while its counterpart, *Din,* may be taken as God's rigor and power of negation, sometimes translated by the word *judgment.*

It is interesting to try, as many have, to find a relationship between love in the human sphere and this universal energy of creativity which is manifested in the variety and abundance of nature and in the force of sex in all living beings. The difficulty arises, however, when one tries to identify the manifestation of this force that is *specifically* human, as distinguished from the sexuality that connects man to animal life. The difficulty is compounded when one understands love as a form of help.

Plato, we may recall, understood by the word *eros* a striving for a new creation through the participation in something more fundamentally beautiful and real. Christian theology, on the other hand, has seen love from the side of God—as a divine outpouring and downflow, much in the way Buddhism speaks of compassion.

And then there is that most mysterious of ideas: God *is* love.

How are we to untangle all this? How can we begin to approach the idea that love is a property of reality, whether we call that reality God or the great universe? Unless we find some way into this idea we shall inevitably remain stranded with the sundry modern psychological perspectives on love: as something which affirms the ego or gratifies sexual need.

One of the difficulties we meet is our tendency to think of the universe in terms of entities such as living things, planets, atoms and the laws that describe them. In that universe we may doubtless find wondrous examples of the fecundity and beauty of natural creation. In the environment around us we may see, without sentimentality, how death and life go hand in hand and how the force of sexual creation and environmental adaptation call into being and sustain a million forms of life.

Illusions About the Force of Sex

But what help does the universe, seen in this way, offer to man? It sustains our bodily existence, yes, but it irrevocably destroys it as well. One need not be a cynic to conclude that the universe perceived through the senses or through scientific presuppositions is more the destroyer than the preserver of man. As for the force of sex, who can say that in our ordinary lives we are much more than its slave? Anyone who thinks that his sexual delights and pleasures are signs of his freedom with respect to this force has never tried to put that freedom to any meaningful test. I am not for a moment connecting these thoughts with the hypocrisies and contradictions surrounding the conventional Judeo-Christian attitude toward sex. I am speaking only of the seemingly unbreakable bonds that connect our attractions and repulsions toward sex with our egoistic emotions and illusions about ourselves. To say that man is a slave of sex is to say that, as he is, the force of sex is one of the most powerful chains that bind him to the ego. And whatever service modern psychology has done by liberating us from the tyranny of sexual guilt is surely counterbalanced by its having led us more and more to define ourselves, both directly and implicitly, in terms of sexual pleasure.

It is true that in many of the great traditions sex openly occupies a central and positive place. But it is particularly useless for us to take this out of context, citing statements or admiring works of art that seem to deify sex. I do not think we have a very clear idea what sort of sexuality is being spoken about or illustrated in these cases. The assumption that traditional civilizations unconsciously understood or experienced sex in the same way we moderns do has supported many nonsensical psychoanalytic and psychological efforts to justify psychological pragmatism.

The main point is that these teachings about sexuality occur within the framework of a great system of ideas—such as the system of Mahayana Buddhism or Samkhya. Without the ability to understand such systems of ideas, how is it possible for us to apply their views about sex to our own lives?

A Definition of Love

Returning now to our central question, we ask: Is the universe made up only of what we perceive through the senses? If so, we may conclude that in it there is no such thing as real help for man.

But have we not all along been dealing with realities that are also part of the universe? Are not *ideas* also part of the universe— ideas, great teachings and great masters?

But we rarely think of the universe in quite this way. In asking what help there is in the universe for ourselves, we tend to include along with ourselves all of mankind. Seeing that our own thoughts and concepts bring us no light, we tend to include with them all ideas. But our whole study has been based on the assumption of the vast real difference that exists between ordinary thought and sacred ideas.

We arrive at the conclusion that for us *cosmic love exists in the form of ideas and human beings who embody, communicate these ideas.*

But, saying this, we come immediately to another conclusion that strikes even closer to home. In our present condition, as we are, we are unable to receive these ideas or hear these communications, nor are we able to recognize which individuals embody them. In short, in our present state of consciousness we are not *able* to receive the help which is offered to us.

Therefore, we may finally define love in the most accurate way for ourselves: Love is that which helps us to understand the truth about ourselves and our possibilities. Are there people and teachings, methods and disciplines, ideas, forms of organization and community pattern, attitudes toward death, sex, material goods, children, time and every other concern of human life which can awaken in us an ableness to hear and understand sacred ideas?

In the universe, as we ordinarily perceive it, man is merely part of planetary nature. Whatever love that exists in such a universe, whatever help for existence, is directed not to humanity alone, but to the whole of planetary nature. In that whole, man is a being who lives and dies as part of the cycle of mutual killing and feeding; he has his place there, but it is the place of a slave among slaves. And it is that universe upon which modern psychology has fashioned its theories of human happiness. Closed to the help that is possible for us as human beings, we receive only the help that is offered to planetary nature.

We need to reiterate: Religion fails us by assuming that we have the ability to understand and take in the great teachings of the past. It fails by assuming that as we are, we are able to receive the love of which it speaks. Science—especially psychology—is a reaction to this blindness of modern religion, a reaction that takes us far over to the other direction by assuming that there is no help for man "out there." In moving to that extreme, in denying the existence of higher levels of intelligence in the external universe, it never finds its way to search for these same levels within man himself, levels of receptivity, organs of intelligence which can mirror in microcosm the laws of a conscious universe. It seems to me that the error of humanism— the error of modern psychology—is not that it seeks for help within man alone, but that it so radically underestimates just what it is that can be found within man. But to glimpse such high possibilities in ourselves it is necessary to have a system of cosmology, a metaphysics, that communicates the existence in reality of these levels. Otherwise, we are stranded with the undeniable fact that, as we are, we are totally cut off from the great cosmos. Psychology underestimates what we can become; modern religion overestimates what we are.

A BRIEF NOTE ON JUNGIANISM

There is currently a revival of interest in Jungianism, understood as a psychological system that makes room for religious man. Carl Jung was the first of Freud's early followers who repudiated scientism in the study of the mind and who sought for categories which were not derived from the physical or biological sciences. He is famous, of course, for his concept of the collective unconscious—the ancient, ancestral psychic life lying beneath the surface of the individual ego. Working with this concept, he and his followers look upon the symbolism of spiritual teachings such as alchemy and Mahayana Buddhism, as well as upon the myths and legends of all nations, as forms by which the collective unconscious expresses itself—even as it struggles to express itself in the dreams and fantasies of every individual.

Nowhere in modern psychology is there more preoccupation with the fabulous contents of the mind than in the school of Jungianism. Indeed, the main thrust of Jungian therapy is the evocation of dreams and the establishment of a positive emotional relationship between the patient and his dream images, which represent, according to the Jungians, the dark powers of life that need to be integrated with the rational, empirical self. The patient is urged to become more and more familiar with this "inner world," to participate in it, to enter into its imagery and mood, to understand, play with, live with, carry and become familiar with it. In this process, man finds his "soul," his inner life.

In the writings of the Jungians, it is difficult to find a clear expression of how they see their relationship to the search for transformation. Sometimes they speak as though they considered their form of psychoanalysis to be a modern, Western version of the Eastern disciplines—and in some respects even an improvement upon them, as though Jungian theories had finally provided a rational expression of what ancient teachings grasped dimly and instinctively. Elsewhere, they more modestly claim only to prepare the psyche for spiritual experience. A leading Jungian analyst asserts that traditional disciplines such as alchemy, Yoga, contemplative Christianity, Sufism and Zen Buddhism are actually concerned with the deepest sense of what Jungians term "personality development."[3]

The career of Jungian psychoanalysis and its ambiguous influence make plain how difficult it is in the modern world to reclaim a sense of inwardness while holding fast to a scientific picture of the universe. The symbols of ancient traditions take on an entirely new meaning when one entertains the idea of a conscious universe containing levels of intelligence that far exceed the highest states experienced by the ordinary, "healthy" mind. If sacred symbols are understood to be the communication by a higher consciousness, they can no longer be approached in Jungian fashion as a bridge between the rational ego and the forces of the unconscious psyche. A spiritual symbol is the expression of an idea, not the result of the play of psychic forces.

The metaphysics of scientism encourages man to stop his search for inwardness at the level of psychic contents. The reason for this is that science does not offer laws of reality that are themselves symbols, that is, the expression of great consciousness. So man has no motive to look within himself for laws, but only for feelings, images or thoughts. From the perspective of science, it is a form of slavery to live according to cosmic laws. From the perspective of esotericism, the possibility of living according to truly cosmic laws is the only meaning of human freedom. And that because the laws of a conscious universe are the patterns by which will maintains itself against resistance, a process mirrored to some small extent in great artistic work.

Microcosmic man has therefore been understood traditionally as man in whom the fundamental laws of the cosmos operate without deviation or distortion. In religious language, this goal has been expressed as "service to God."

From this perspective, then, the transformation of human nature begins as the individual acquires the sensitivity and therefore the power to live under higher and more universal laws, laws which govern the great cosmic movement between levels of being. No psychology that does not effectively recognize a universe of levels and conscious hierarchy can possibly establish a relationship to the ancient teachings, far less prepare an individual for the work of spiritual struggle. Surely, a sacred psychology is a teaching that can guide men and women into the study of these laws as they operate within the structure of human nature. And, surely, a spiritual master is an individual who, having understood these laws, can guide another past the obstacles that prevent his living in accordance with them.

Therefore, the question still remains open whether any psychiatrist, even a Jungian, can prepare an individual—except by accident—for the search for consciousness. In fact, as has been suggested, there are grounds for wondering if perhaps modern psychotherapy works in precisely the opposite direction. I offer the following quotation from the previously cited writings of Sri Anirvan as food for thought on this issue:

> Psychoanalysis works on the intensive exposure of the emotions. They are brought out into the daylight with the object of cleaning out the subconscious, and thus are lived over again, which means that each of them is amplified. Instead of belonging only to the damaged part of the being, they invade the whole field of *prakriti,*[†] as the tares spread over a wheatfield.
>
> The very structure of emotion must be denied as such, for it arises from one's subjective view of external elements. . . . The traditional techniques pay no attention to emotion. Even the Vaishnavites, who worship Krishna and make full use of emotion in their fundamental attitude of adoration, use only those sublimated elements of emotion that sustain the ideal.
>
> Several levels have to be passed through before one can know how to minimize emotion at *its very source,* to know it, to isolate and master it, and finally to be able to get rid of it. The lower stage is to realize once and for all that emotions are a debt to be paid; this realization is the beginning of a process that uproots them. This is the process of eradication. The second stage is to conceive of emotion as a surrender, an always recurring pattern. It is part of the automatism which becomes evident and really does not exist. The third stage is that in which a "light form" of an ideal is methodically and voluntarily put in the place of the "heavy form" of an emotion which is oppressive. . . .

[†]Essential to the Samkhya is the distinction between *Prakriti* and *Purusha. Prakriti* may be translated as "Great Nature" and comprises the entire universe of things, laws and movement. *Purusha* is Great Consciousness which pervades Prakriti at every point and is yet completely distinct from it. The Samkhya system deals with these two realities as they operate in the universe and in the microcosm and is a method of awakening the prakriti in man to the influence of consciousness through a discipline of presence-to-oneself in the midst of the ordinary conditions of life.

Renunciation *(vairâgya)* is the voluntary giving up of all emotions whatsoever. This notion, supported by a long tradition, goes hand in hand with life. . . . To be capable of mastering an emotion, one has first to evaluate it and dispose of it for what it really is—the distortion of an uncontrolled and misplaced sensation.

When the intestines are out of order, one follows a strict diet. The cure comes about by abstaining for the time being from certain foods. Thus the body regains strength. Psychically, power is gathered. This method is the exact opposite of psychoanalysis, which digs about in the ego. *Samkhya* places you under a cosmic force and disregards the ego, saying, "Why are you afraid of this or that? All these things are only movements of *prakriti,* aspects of the eternal recurrence in men, animals, the whole of nature." One must learn to live in the very movement that shapes and molds *prakriti,* without trying to escape from it. To look at *prakriti* and see her agitated movement makes it possible not to identify with her. I observe what she is. By doing that, I feel the movement in myself, but I do not linger on the fact that I was created from the same matter. Time plays an important part in this discipline, and also patience. On the part of the Guru this patience is pure love.

Emotion does not enter into any spiritual discipline, because in itself emotion has no consistency. It is only a movement of *prakriti.* When the mind is perfectly calm it is like the still water of a mountain lake. The slightest ripple on the surface is an emotion.

What happens to it? If *Purusha* allows this ripple, however slight, to intensify and become a wave, he himself will be swallowed. Blind emotion is then master of the situation, although in fact it has no *raison d'être.*

If this emotion, while it is still only a ripple, is voluntarily interiorized, then little by little, because of its inconsistency, it will disintegrate and of itself go back whence it came.[4]

THE THIRD BROTHER

In conclusion, let us return to our tale of the two youths meeting on the side of a mountain. Were I to attempt to reconstruct the whole story from this tiny fragment, I would work on the assumption that it is a tale about the nature of love, as it is spoken of in symbol and legend in all the traditions of mankind. I suspect that the two young men are introduced as brothers, as two aspects of everyman's human nature, and that the story is meant to warn us against identifying ourselves solely with either the inward or the outward movement of energy within the psyche. I see the narrative following the crane now flying free, but with the long golden thread trailing from its neck and brushing against the earth from time to time. And I imagine as well that we are given an unmistakable picture of the fate of the second brother being dragged with ever-increasing violence down the mountain, finally stumbling and falling injured as the black dog breaks loose. As in many such fairy tales, there is probably a third brother in this story, younger than the first two, but more sincere in his ways and more open to deep learning about the totality of himself. He represents the embryonic human soul in its ancient meaning as that in ourselves which can stand between the highest and the lowest within us and which, for us to live and evolve, must care for both the animal and the spiritual in man. Long ago, Western religion turned away from this "third brother" and gratuitously insisted that all human beings were born with a fully developed "immortal soul." As a result, teachings which recognized that long and exacting work was needed to develop the real *I* in man that stands "between the worlds" went underground and kept their secrets well. The subtle discipline through which an individual learns impartially to take in impressions of both sides of his nature became for the West a forgotten science.

I see the third brother grasping hold of the golden thread just before the crane soars out of reach and I see him then laying hold of the powerful black dog. There he stands, connected for the moment to both the spiritual teachings and the natural desires of the body and the social self which must be neither indulged nor ignored, but *known* if man would acquire the energy to actually live the truths which the traditions communicate. And so he proceeds with the thread in one hand and the iron chain in the other. Who will help him? Is there a love for man, for ourselves, which can care from moment to moment for the growth of both sides of our nature, the part that hears the whispering from the cosmos and the part that lives and serves the biological life of earth?

NOTES

1 Robert E. Ornstein, *The Psychology of Consciousness,* San Francisco,W.H. Freeman and Company, 1972, pp. 96—99.

2 Sigmund Freud, *The Future of an Illusion,* translated by W. D. Robson-Scott, Garden City, New York, Doubleday, Anchor Books, pp. 21—23.

3 James Hillman, *Insearch,* New York, Charles Scribner's Sons, 1967, p. 84.

4 Lizelle Reymond, To *Live Within,* New York, Penguin Books, Inc., pp. 204—207

Chapter Six

Magic

There are many legends of magicians who invoked a spirit but lacked the power to control it. This happened either because the magician forgot the words of the invocation, or in some way broke the magic ritual, or because he invoked a spirit stronger than himself, stronger than all his invocations and magic figures.

All these instances, of the men who break the ritual of initiation in the Mysteries, or of the magicians who invoke spirits stronger than themselves, equally represent, in allegorical form, the position of a man in relation to new ideas which are too strong for him and which he cannot handle because he has not the required preparation. The same idea was expressed in the legends and tales of the sacred fire which consumed the uninitiated who incautiously approached it, and in the myths of gods and goddesses the sight of whom was not permitted to mortals, who perished if they looked upon them. The light of certain ideas is too strong for man's eyes, especially when he sees it for the first time. Moses could not look at the burning bush; on Mount Sinai he could not look upon the face of God. All these allegories express one and the same thought, that of the terrible power and danger of new ideas which appear unexpectedly.

The Sphinx with its riddle expressed the same idea. It devoured those who approached it and could not solve the riddle. The allegory of the Sphinx means that there are questions of a certain order which man must not approach unless he knows how to answer them. Having once come into contact with certain ideas man is unable to live as he lived before; he must either go further or perish under a burden which is too heavy for him.

P.D. Ouspensky

Magic and Sacred Psychology

T he universe of scientism is a world devoid of consciousness and pur-
posefulness. And the various sciences describe the laws that govern
reality when we see it in this limited way. In a multileveled universe
we must accept that this level or spectrum of reality also has its existence —
just as it is possible to describe a great painting such as the Mona Lisa solely
in terms of the laws of chemistry and light refraction without considering
even that it is a picture of a woman and without intuiting the intention of the
artist which led him to arrange the pigments in this way and no other. Laws
exist at all levels. And, at their level, the laws of mechanism (in the modern
sense of the word) are just as valid and rigorous as the laws of consciousness. Error enters only when we indiscriminately assume there are no other
levels than the one we see.

That is why modern psychology stands out among the sciences as a sort
of strange disfigurement. The whole enterprise of modern, scientific psychology is rooted in an impossible contradiction: the attempt to subsume
one level of reality under laws that govern a lower level. In attempting this,
psychology has only succeeded in bringing down the idea of consciousness
to the idea of automatism (patterned in an earlier time after the model of the
machines of industrialism and in present times after the model of the computer). In this sense, as we shall see, behaviorism, although it is in a sense the
farthest from a sacred psychology, is in its own way the most valid branch of
modern psychology. In its pure forms, it never pretends to speak about consciousness.

Psychology, properly so-called, must therefore always be a sacred science—in the sense that it is by definition the science of the possibility of
man living his life in conscious accord with fundamental causal laws. Since
ancient times, and among all peoples, the practice and theory of magic has
existed as a sacred psychology in this sense.

Historians, theologians and anthropologists have tended to draw a sharp
line between the idea of man as a magician and man as a saint—a distinction
between magic and wisdom, between power and goodness as qualities of
developed man. Usually, this distinction is made to the detriment of the

"magician." The implication is that extraordinary powers can exist in an evil or destructive human being. I think there is an important truth in this distinction. But I do not believe it will take us very far unless we consider that magic and religion are two different languages that deal with the same perennial theme of man's conscious evolution. Magic concerns the development of man through the growth of will, while the path of religion on the whole deals with the development of man through the denial of willfulness, egoistic will. The religious path says to man: "Surrender your trifling sense of ability which you magnify into something fantastic and grotesque, and allow the power of God to operate through you." The path of magic says: "Create in yourself a will and an individuality that is an instrument of higher cosmic energies." Both magic and the path of religion in their authenticity, and as properly understood, deal with the transformation of man into a being who can consciously manifest fundamental power (or the "Will of God").

As a branch of modern science, psychology cannot recognize the idea of higher causality in the universe. Thus, it is least hypocritical in the form of behaviorism which treats man as an automatism operating under the same laws as the objects of modern physics and biology. And, like the cosmos, man surely has within his mind functions that are automatic in this sense. It is even possible to define "fallen man" (in the language of religion) as man who lives, thinks, desires and acts according to the laws that govern these lower or automatic aspects of his being. Thus, in principle, behaviorism is the true science of fallen or automatic man. The error of behaviorism is that it believes this is all man can ever be. At the same time, and for reasons we shall go into presently, it radically underestimates the complexity of the human automatism even in its pathological state. And for this reason, because it does not see the real failure of the human mechanism, behaviorism is powerless to assuage human suffering—even apart from the fact that it is blind to the possibilities of man's conscious evolution.

The Practice and Language of Magic

It is customary to look upon the growth of modern science as representing an increase in man's power over the forces of nature. The scientist is regarded as being able to do many things that would have seemed "magical" to previous cultures.

The classic example of this concerns alchemy. One still hears it said that the development of nuclear physics allows modern man to realize the dream of the alchemists. The medieval alchemists are almost always pictured either as charlatans or as pathetic fantasists trying to get rich by transmuting baser metals into gold.

And, certainly, if one reads the texts of the alchemists literally, without symbolic understanding, that is exactly what they seem to be talking about. Yet the moment one is free, if only slightly, from the literal mind, one begins to see that alchemy is a form of the language of magic—an expression, that is, of the ideas and methods of sacred psychology. A particularly clear contemporary statement of this understanding of alchemy is to be found in the study by Titus Burckhardt, to which we have already referred in previous chapters:

> In fact alchemy may be called the art of the transformation of the soul. In saying this I am not seeking to deny that alchemists also knew and practised metallurgical procedures such as the purification and alloying of metals; their real work, however, for which all these procedures were merely the outward supports or 'operational' symbols, was the transmutation of the soul. . . Thus, in contradistinction from the usual reproach against them, the alchemists did not seek, by means of secretly conserved formulas in which only they believed, to make gold from ordinary metals. 'Whoever really wished to attempt this belonged to the so-called 'charcoal burners' who, without any connection with the living alchemical tradition, and purely on the basis of a study of the texts which they could only understand in a literal sense, sought to achieve the 'great work.'

> Alchemy. . . is primarily neither theological. . . nor ethical; it looks on the play of powers of the soul from a purely cosmological point of view, and treats the soul as a 'substance' which has to be purified, dissolved and crystallized anew. . . Its spiritual, and in a certain sense, contemplative nature resides directly in its concrete form, in the analogy between the mineral realm and that of the soul; for this similarity can only be perceived by a vision which can look on material things qualitatively—inwardly, in a certain sense—and which grasps

the things of the soul 'materially'—that is to say objectively and concretely. . . With its 'impersonal' way of looking at the world of the soul, alchemy stands in closer relation to the 'way of knowledge' (gnosis) than to the 'way of love.' For it is the prerogative of gnosis—in the true and not the heretical sense of the expression—to regard the 'I'—bound soul 'objectively,' instead of merely experiencing it subjectively.[1]

Now, if we in the modern age have so wildly misunderstood the meaning of alchemy, from which we are separated by only a few hundred years of time, how much more may we have misunderstood similar systems of even more remote times and places? We confidently sweep in whole teachings and cultures under the label of "magic," thereby relegating them to the level of superstition. We assume that all peoples meant exactly what we mean by "power over nature."

But, following the lines of thought we have been developing in this book, we may define power as the experiential and practical knowledge of real causes. And we may define slavery as bondage to the effects of causes. The principal ingredient of psychological slavery to nature is therefore the inability to distinguish effects from causes. Compared to this inability, ordinary error is secondary. The fundamental aspect of our ignorance is that we mistake effects for causes and hence remain blind to real causality in its connection to cosmic law. But it must be added that, as we are told, real causes cannot be known through ordinary thought or external perception alone, but must be studied directly through disciplined inner experience.

I take this study to be the meaning of the practice and language of magic. Of course, by magic I do not mean to include superstitions that are the residue of great systems of magic. It is reasonable to suppose that if alchemy spawned countless "charcoal burners" who took its teachings literally and attempted to apply them for egoistic purposes, then so did other systems of magic in other cultures and in times past. How much of what we call magic is really only these residues of superstition compounded by suggestion and our own fantasies?

Magic

The Laws of Power

I am suggesting that microcosmic man is a being who lives in the world of real causes, a being in whom the great causal energies of the universe move and develop according to patterns that shape the cosmos itself. Strictly speaking, this is the very meaning of the word man: a living reflection of total reality. Microcosmic man, perfected man, is therefore an agent of fundamental causal power.

If that is so, then what about ourselves? We imperfect and ignorant human beings—what are we? What place do we occupy in the scheme of reality? One must begin by answering that as he is, man on earth is only an effect. Everything that we see in ourselves—all our thoughts, feelings, sensations and actions—are only effects, results of unknown causal powers. Like the contents of the mind, the contents of our very lives are the expression of laws which we do not have the means to comprehend.

Surely the enormous difference between esoteric natural science and modern natural science stems from this consideration. Esoteric, or sacred science, is the study of the movement of force between the Creator and his creation, between the pole of consciousness, unity and intention on the one hand, and the pole of manifestation, articulation and automatism on the other. Such a science may also be called the study of the laws of power— in that all genuine creation exhibits lawful, periodic fluctuations with respect to the ingathering and outflow of energy. The aim of this study is the mastery of these laws of power first in our own inner life and then with respect to the world around us. Such is the spiritual task and possibility of man as understood in the language of magic. Born to be the "lord of creation," a human being becomes such by first becoming lord of himself. Only then can he be an instrument of fundamental causality.

Anthropologists who study magic in primitive cultures tend to be fascinated by external marvels (such as "fire-walking") and remain blind to the inner work of gathering psychic energy which lies at the heart of these rites—at least in cultures which have kept the whole meaning of their ritualistic tradition alive. As a rule, we have no wherewithal to distinguish those cultures that have preserved the inner aspect of their teaching from those in which only the external aspects have remained in the form of superstition and self—hypnosis. In these latter cultures there perhaps exist "inexplicable" and apparently miraculous happenings which are engaged in blindly and uncomprehendingly. Who knows what forces are gathered around a people whose conditions of life make extraordinary physical and emotional demands on them? But magic, properly so—called, only exists when these forces are precisely understood and mastered in oneself. Possibly what is called "black magic" exists when this inner work is forgotten, while the conditions of life remain extraordinary and therefore a magnet for forces which we in the comfortable conditions of modern life can only fantasize about or look upon with an indulgent sense of superiority.

The Process of Deception

Was it only my adolescent fantasies that drove me, for quite some while, to study "magic" with the same eagerness that I studied science? Or was I also responding—in a garbled, childish fashion, to be sure—to some deeper dream that has defined the aspiration of Western man himself? I do not think we can so easily set aside the observations of countless historians and philosophers that the genius of Western man lies in his vision of miraculous power and great action. Of course, looking back on my own early experiments with "magic" it is obvious that I never came near to contacting a superior causality. That possibility was never even in question since all I ever dealt with were the tricks of the stage magician. Yet I cannot but feel that in my dreams of magic there were echoes of an intensely serious purpose, a possibility for man that needs to be much better understood in the modern world. The teachings of the East come to us with a call for inwardness, and yet we in the West respond as creatures born for action in the world. It seems we cannot do otherwise. Perhaps we ought not do otherwise.

As an adolescent, I desired the magical—something so new, a reality so vibrant and free, and so miraculous as to dissolve all fear of life. Of course, I realized that all I would learn would be the methods of producing illusions—imitation magic and imitation power. Only now, looking back on those years, do I begin to see why I was drawn, and do I begin to appreciate what I was taught.

It was a club for young magicians called, appropriately, The Sorcerer's Apprentices. Once a week we met in a dingy office building to go through our routines and criticize each other. Our leader was an elegant, flamboyant young man who to many of us was far more than a mere instructor of techniques. He seemed a representative of the miraculous and it was his influence that kept the dream of magic alive even as we were all learning the mechanics of mere trickery.

He used also to hold regular meetings in his apartment to which I was invited not long after joining the club. These meetings were attended by the club members—though not by all of them—and also by several other adults who had no interest whatever in stage magic. The purpose of these meetings was to discuss ideas about philosophy, especially the occult sciences, spiritualism and life after death. I was by far the youngest person present at these meetings and throughout the whole year I listened to the discussions with bewilderment and skepticism, hardly ever speaking or offering an opinion.

In those days such questions were unfashionable in intellectual circles. Nowadays, of course, that is not so true. Even sober, responsible scientists are now thinking seriously about extrasensory perception, prophecy, psychokinesis, spiritual healing and so forth. . .

In any event, during the course of that year I mastered a wide spectrum of magic tricks. I became quite adept at sleight of hand with cards and coins, I learned many secrets of what is called in the trade "mental magic"—mind-reading effects, predictions, all of which interested me enormously—and I worked as much as I could with the astounding variety of mechanical appliances that are the tools of professional magicians: elegant silver tubes and boxes in which all manner of objects could be made to appear and disappear, ingenious and mysterious-looking containers, bowls, bells, tripods, pitchers, as well as all the invisible gimmicks which the public never sees or hears of and by means of which things are produced out of thin air, float above the ground and are magically joined and severed. Finally, I was able to ferret out the secrets of almost all of the great stage illusions: levitations, severed bodies and the incomparable feats of Houdini.

Yet all this knowledge did not make me skeptical about the possibility of real magic. On the contrary. Even when I knew exactly how an effect was produced, even when I was doing it myself, most of these tricks only reawakened the dream of magic. I could never watch or perform a trick that was being done with serious overtones without feeling the sense of something new and strange like the signature of another world.

At the same time, my amazement never dimmed concerning the ease and simplicity by which people, myself included, could be made to see something that did not happen, or to not see something that was taking place right in front of them. After a while, I carried this impression around with me constantly, and it has never quite left me—I mean the possibility that everything we so confidently "see" is also an illusion. Not all of my young colleagues shared this sort of impression; for many of them the study of trickery made no mark on their belief in the so-called "real world" of the senses. But for me the result of studying magic was the recurrent feeling that if I could be tricked so easily into seeing a solid pencil pass through a sheet of glass, then the whole world in front of me could be the product of "trickery."

I cannot convey the feelings of fear and astonishment that were evoked in me when I actually witnessed the full process of deception taking place in others and in myself.

Passive Attention

Now the main instrument for producing this deception is a technique known as "misdirection." Simply stated, misdirection is the art of attracting someone's attention so that he looks where you want him to look and, as a result, sees what you want him to see. Without this, the art of stage magic would not be possible. Almost every trick, every illusion great and small, depends on the passivity, the weakness, of human attention.

I do not claim that this is a new discovery. Everyone will acknowledge that perception can be influenced through the distraction of attention. But how many grasp the importance or the pervasiveness of this phenomenon? Having witnessed many occasions when my own mind constructed and believed in a totally fictitious event due to the passivity of attention, I cannot imagine why this phenomenon is not first on everyone's list of things to be investigated, studied and changed.

I have read many studies, including the exciting books of Carlos Castaneda, concerning perceptual experience and the way it is affected by conditioning and social context. None speaks at all about what seems to be the fundamental fact: man's passive attention. They are all concerned with the effects of this passivity upon the perceiver—what I have called the contents of the mind—and not with the structural aspect of human nature which allows these effects to take place.

Professional magicians work at misdirection in many ways. In most cases, all that is needed is that the magician look where he wants you to look. Your eyes will follow his, and your perceptual associations will automatically create the illusion he seeks. On the whole, we like to be fooled this way. We assume, of course, that being fooled like this is a relatively rare event which only takes place in magic shows or at the hands of unscrupulous people such as crooked gamblers. But suppose it is happening all the time, without any magician having to be present? If it is, then all talk about freedom of the will or the acquisition by man of power is totally meaningless.

Visual misdirection is only one method. Usually, the magician will tell you in advance what you are going to see. And because most of us enjoy magic and like to be fooled in this entertaining way, we allow the magician's words to evoke in us the corresponding perceptual associations. A complete skeptic, or someone who doesn't want to be fooled, will generally see only puzzles to be figured out rather than magical effects, but even he will have to watch over his attention.

However, the explicit mention of what is going to take place is only one part—and not the most important part—of what we might call mental misdirection. Much more important is the role of context, custom and unspoken rules. To take a very elementary example, when the magician pours milk from a pitcher into a paper cornucopia he does not have to tell me that the pitcher is made of ordinary glass. On the contrary, he does his best not to call my attention to the pitcher at all (in this trick, the pitcher has a plastic lining that contains the milk, which only seems to be flowing into the cone) . If the magician is skillful, it never occurs to me that the pitcher is doctored. I assume it is ordinary. But what is an assumption? It is a product of passive attention. And as the result of such an assumption, the mind and the perceptual apparatus construct an event. As I have said, this is a relatively simple example—the situation can become very complex and subtle, and even the cleverest—most scientific—person can easily be fooled. For no amount of what we call intellectual brilliance can compensate for passivity of attention. Attention and thought are two very different functions—a fact which is not recognized or appreciated in modern psychology.

The modern neglect of the study of one's own attention raises important questions, to say the least, about the ultimate validity of scientific knowledge. What sort of universe is accessible to men and women who have little or no control over their attention? And who do not even suspect the extraordinary role that attention plays in the process of their own cognition, perception and understanding? What is the relationship between the quality of attention and what we have been calling states of consciousness? Is a mechanistic picture of the universe a reflection of our mechanical, passive attention? In any event, we have here one clue as to why the persistent study and perfection of attention have played so dominant a role in the psychological disciplines of the ancient spiritual traditions.

So far we have been speaking of passive attention with respect to perceptual focus and conceptual processes—visual and mental misdirection. But there is yet a third form of misdirection which the professional magician instinctively uses: emotional misdirection. I have already spoken about this when referring to the desire to be fooled and, in my own case, when referring to the feeling of another world, another reality.

Suggestibility

There is no end to the kinds of emotion that will help the magician. If he knows his business at all, he will immediately sense whether the spectator's emotional state is working for or against him. Many of the great stage magicians owe their success mainly to an instinctive talent for inducing a certain emotional set in the entire audience.There are many emotions that enter into the process of deception. I myself—without really understanding what I was doing—often made use of fear in order to cause the spectator to think and see something that did not happen. I mean the fear people have of being thought stupid, as well as their excessive desire to please When spreading out a deck of playing cards and "forcing" one particular card upon the spectator, I have seen him hesitate for a split second as the perception of what is really taking place crosses his mind like a swiftly moving shadow. Yet immediately after, he will eagerly "select" the intended card which I all but put in his hand for him. And not one person in a hundred is aware of what has happened. Almost everyone is quite sure he has "freely" chosen a card at random.

All of the above (which would need a long essay to describe in full) and much, much else besides, make up the phenomenon of suggestibility. Suggestibility is the sum-total product of man's passive attention, and all stage magic is based on suggestion. The spectator is simply not aware of the way his perceptions, thoughts and emotions automatically gather together to construct illusory objects and events. In everyday life we say, "I saw it with my own eyes, I actually experienced it," in order to lay aside any suspicion that we have been deceived. But say this to any seasoned stage magician and it will only bring a smile to his lips.

By no means did I suspect the significance of all this when I was a teenager first learning the techniques of trickery, which most people take as a pleasant, but trivial pastime. As I have said, all I felt was the intimation that the world around me might also be an illusion.

There was a profoundly different "taste" to the moments when I actually witnessed the process of deception taking place in myself and others. Only gradually, over the years, have I begun to realize that without moments of directly "tasting" self-deception, it is impossible to understand the condition of man, and why his life goes the way it does.

For it would surely be naive to believe that our attention is passive only when we are in the hands of a stage hypnotist or professional magician. Nor does the magician himself necessarily understand the importance of what he is working with. We must conclude that our attention is almost always passive, that we are almost always under the sway of suggestion which is only accentuated and accelerated by the stage performer.

Maya

In my college years it always produced a deep impression on me, not often shared by peers or instructors, when I read in some ancient text that reality is a mental construct, and that the world we so confidently live in, and suffer over, is actually layer upon layer of illusion. But the attempts of most modern scientists, philosophers and psychologists to explain and correct this state of affairs leave me cold because almost nowhere do they discuss the overwhelming role of attention. They always want to improve the situation through new concepts or through stricter canons of experimentation, as though it were possible to improve man's quality of perception without a radical transformation of the quality of his attention.

Here it is interesting to note that in the days when modern psychology was struggling to establish its credentials as a science, the behaviorists shocked many people by disclaiming the use of introspection. As is usually the case, two opposing schools of thought arose: one making use of self-examination, the other, the behaviorists, relying only on external observation. Along with the behaviorists, many philosophers began arguing that a man has no privileged access to his own psyche, and the problem of self-knowledge was reduced to the problem of how to make external, accurate observations of behavior. Deprived of the license to attempt self—observation, other philosophers began speaking about the use of language as a key to resolving problems about human nature and our understanding of the universe. All of this together—behaviorism, language analysis or language acceptance (the "ordinary language" school, as it is called)—has had enormous influence on the intellectual climate of America. Many brilliant and famous men such as Ludwig Wittgenstein and Bertrand Russell have been part of it. Yet it all rests on a failure to distinguish thought from active attention. All that these behaviorists and philosophers have shown is that it is impossible to know the mind through thought associations. But self-knowl-

Magic

edge, in the sense communicated by Socrates, or the great Christian contemplatives or the Eastern masters, is concerned with the development in man of a different quality of attention, and has nothing to do with what we generally call introspection. Here again, the modern scientific temper—like the child in the fable—saw that the emperor had no clothes, but neglected to understand that he was nevertheless still the emperor. What we proudly call self-knowledge (through introspection) is surely a form of associative thinking and self-indulgence made possible by the passivity of our attention. But to conclude from this that direct self-knowledge is impossible is equally naive.

The difficulty is that the modes and gradations of human attention cannot be argued about with logic or concepts, but have to be experienced. It is indeed hard for us to accept that we have never experienced anything more than a fleeting glimpse of a freer level of attention. And because of our world view, even these rare and fleeting glimpses do not raise serious questions in our minds about the possible transformation of the psyche. We never ask: What is this life force called attention? From where in the universe and from where in ourselves does it come, and what changes in our being would result were we to work for more access to this force?

A legend tells how once Narada said to Krishna, 'Lord show me Maya.' A few days passed away, and Krishna asked Narada to make a trip with him towards a desert, and after walking for several miles, Krishna said, 'Narada, I am thirsty; can you fetch some water for me?' 'I will go at once, sir, and get you water.' So Narada went. At a little distance there was a village; he entered the village in search of water, and knocked at a door, which was opened by a most beautiful young girl. At the sight of her he immediately forgot that his Master was waiting for water, perhaps dying for the want of it. He forgot everything, and began to talk with the girl. All that day he did not return to his Master. The next day, he was again at the house, talking to the girl. That talk ripened into love; he asked the father for the daughter, and they were married, and lived there and had children. Thus twelve years passed. His father-in-law died, he inherited his property. He lived, as he seemed to think, a very happy life with his wife and children, his fields and his cattle, and so forth. Then came a flood. One night the river rose until it overflowed its banks and flooded the whole village. Houses fell, men and animals were swept away and drowned, and everything was floating in the rush of the stream. Narada had to escape. With one hand he held his wife, and with the other two of his children; another child was on his shoulders, and he was trying to ford this tremendous flood. After a few steps he found the current was too strong, and the child on his shoulders fell and was borne away. A cry of despair came from Narada. In trying to save that child, he lost his grasp upon the others, and they also were lost. At last his wife, whom he clasped with all his might, was torn away by the current, and he was thrown on the bank, weeping and wailing in bitter lamentation. Behind him there came a gentle voice, 'My child, where is the water? You went to fetch a pitcher of water, and I am waiting for you; you have been gone for quite half an hour.' 'Half an hour!' Narada exclaimed. Twelve whole years had passed through his mind, and all these scenes had happened in half an hour! All this is Maya. In one form or another, we are all in it. It is a most difficult and intricate state of things to understand. It has been preached in every country, taught everywhere, but only believed in by a few, because until we get the experiences ourselves we cannot believe in it. What does it show? Something very terrible[2]. . .

Attention to Oneself

Having brought our discussion to this question and having suggested that attention is the key to magic, both as deception and as real power, we are in a position to turn to the great masters for instruction. Only we shall have to be very mobile in our approach, recognizing that the name for attention in its many aspects varies considerably not only from tradition to tradition, but even within the writings of the masters themselves (for example, sometimes it is associated with the words "mind" or "intellect"—or sometimes with "light," or even "spirit"). We need to remember that we are inquiring into something of which we ourselves have very little experience, and that even with the best of motives we will barely be able to see behind the surface of this issue which has such extraordinary implications about both the nature of man and the living structure of the universe.

Perhaps the best beginning is to recall that we are speaking about a sustained attention directed upon the processes in oneself (thoughts, sensations, emotions) as they occur as part of ordinary experience. In the

Buddhist tradition this effort is termed mindfulness and is described at length in the discourses of the Buddha. The following is a brief excerpt from the Satipatthana-sutta, the discourse of "The Setting-up of Mindfulness":

> And how, O priests, does a priest live, as respects the body, observant of the body?
>
> . . . O priests, a priest, in walking thoroughly comprehends his walking, and in standing thoroughly comprehends his standing, and in sitting thoroughly comprehends his sitting, and in lying down thoroughly comprehends his lying down, and in whatever state his body may be thoroughly comprehends that state.
>
> Thus he lives, either in his own person, as respects the body, observant of the body, or both in his own person and in other persons, as respects the body, observant of the body, either observant of origination in the body, or observant of destruction in the body, or observant of both origination and destruction in the body; and the recognition of the body by his intent contemplation is merely to the extent of this knowledge, merely to the ex-tent of this contemplation, and he lives unattached, nor clings to anything in the world.
>
> Thus, O priests, does a priest live, as respects the body, observant of the body.
>
> But again, O priests, a priest, in advancing and retiring has an accurate comprehension of what he does, in looking and gazing has an accurate comprehension of what he does; in drawing in his arm and in stretching out his arm has an accurate comprehension of what he does; in wearing his cloak, his bowl and his robes has an accurate comprehension of what he does; in eating, drinking, chewing, and tasting has an accurate comprehension of what he does; in easing his bowels and his bladder has an accurate comprehension of what he does; in walking, standing, sitting, sleeping, waking, talking, and being silent has an accurate comprehension of what he does.
>
> Thus he lives[3]. . .

This portion of the Satipatthana-sutta goes on, in an identical vein, advising the priest to maintain his attention upon all aspects of his being: sensations (pleasant sensations, unpleasant sensations, indifferent sensations, interested and pleasant sensations, disinterested and pleasant sensations, etc.); emotion (a passionate mind, a mind free from passion, a mind full of hatred, a mind free from hatred, an infatuated mind, a mind free from infatuation an intent mind, a wandering mind, an exalted mind, an inferior mind, a concentrated mind, an unconcentrated mind, an emancipated mind, an unemancipated mind, etc.); dispositions, thoughts, perceptions, reactions, sounds, tastes, colors, odors—in short everything that could possibly be seen as the contents of the mind.

Among the numerous commentaries concerning this way of living, one often comes across the statement that through this work of attention both a knowledge of reality and a new form of energy (or "controlling power," as one text calls it) arise spontaneously.

However, these texts are probably of limited usefulness for those of us not engaged in the full form of the traditional discipline. It seems that not even the traditional teachers of contemporary times fully realize the deadening effect of language which, though psychologically correct from their point of view, has no metaphysical or emotional resonance in the modern listener. These modern teachers seem unaware of how accustomed we are to take ordinary "straightforward" psychological language and single-leveled terminology on the basis of our familiar associations, thus making the great teachings of a Buddha or Patanjali (founder of the Yoga system in Indian philosophy) into something that refers only to our familiar thoughts and feelings rather than to the entire universe in its aspect of consciousness and subtle, large-scale, difficult-of-access benevolent power.

But many of the writings (wrongly taken by us as "poetic") of the ancient and medieval masters have a mysterious ability to simultaneously echo truths about myself and the hidden universe. This can even come through in translations.

Attention and the Evocation of Conscious Energy

Concerning the work of expanded attention and the universal energies which it can attract, I turn to a text by the medieval Jewish sage, Moses Maimonides. The following is from the concluding portions of *The Guide for the Perplexed*, a book written for pupils who wished to take a step beyond the limits of exoteric religion and the literal interpretation of biblical teachings.In reading this selection, it is necessary to set aside our customary understanding of the words "intellect," "Providence," "knowledge" and, even, "God," and remember that we are speaking about attention:

I have shown you that the intellect which emanates from God unto us is the link that joins us to God. You have it in your power to strengthen that bond, if you choose to do so, or to weaken it gradually till it breaks, if you prefer this. It will only become strong when you employ it in the love of God, and seek that love. . .You must know that even if you were the wisest man in respect to the true knowledge of God, you break the bond between you and God whenever you turn entirely your thoughts [attention] to the necessary food or any necessary business; you are then not with God and He is not with you; for that relation between you and Him is actually interrupted in those moments. . .

I will now commence to show you the way how to educate and train yourselves in order to attain that great perfection . . .

Maimonides then outlines a course of disciplining the attention, requiring years of gradual practice both alone in meditation and in the midst of ordinary life situations.

He then continues:

The "Covenant"

When we have acquired a true knowledge of God, and rejoice in that knowledge in such a manner, that whilst speaking with others, or attending to our bodily wants, our mind is all that time with God; when we are with our heart constantly near God, even whilst our body is in the society of men; when we are in that state which the Song on the relation between God and man poetically describes in the following words: "I sleep, but my heart waketh; it is the voice of my beloved that knocketh" (Song v. 2) :—then we have attained not only the height of ordinary prophets, but of Moses, our Teacher, of whom Scripture relates: "And Moses alone shall come near before the Lord," "But as for thee, stand thou here by me" (Deut. v:31). . . .

The Patriarchs likewise attained this degree of perfection. . . Their mind was so identified with the knowledge of God, that he made a lasting covenant with each of them. . .

"Providence" and Accident

We have already stated. . . that Divine Providence watches over every rational being according to the amount of intellect which that being possesses. Those who are perfect in their perception of God, whose mind is never separated from Him, enjoy always the influence of Providence. But those who, perfect in their knowledge of God, turn their mind sometimes away from God, enjoy the presence of Divine Providence only when they meditate on God; when their thoughts are engaged in other matters, Divine Providence departs from them.

. . . This person is then like a trained scribe when he is not writing. Those who have no knowledge of God are like those who are in constant darkness and have never seen the light. . .

Hence it appears to me that it is only in times of such neglect that some of the ordinary evils befall a prophet or a perfect and pious man; and the intensity of the evil is proportional to the duration of those moments, or to the character of the things that thus occupy their mind. . . . When he [man] does not meditate on God, when he is separated from God, then God is also separated from him; for it is only that intellectual link with God that secures the presence of Providence and protection from evil accidents. Hence it may occur that the perfect man is at times not happy, whilst no evil befalls those who are imperfect; in these cases what happens to them is due to chance. This principle I find also expressed in the Law. Comp. "And I will hide my face from them, and they shall be devoured, and many evils and troubles shall befall them; so that they will say in that day, Are not these evils come upon us because our God is not among us?" (Deut. xxxi. 17). It is clear that we ourselves are the cause of this hiding of the face, and that the screen that separates us from God is of our own creation. . . . There is undoubtedly no difference in this regard between one single person and a whole community. It is now clearly established that the cause of our being exposed to chance, and abandoned to destruction like cattle, is to be found in our separation from God.

The "Presence of the King"

Further on, Maimonides writes of attention and inattention through the simile of a man at ease among his familiars:

We do not sit, move, and occupy ourselves when we are alone and at home, in the same manner as we do in the presence of a great king; we speak and open our mouth as we please when we are with the people of our own household and with our relatives, but not so when we are in a royal assembly. If we therefore desire to attain human perfection, and to be truly men of God, we must awake from our sleep, and bear in mind that the great king that is over us, and is always joined to us, is greater than any earthly king, greater than David and Solomon. The king that cleaves to us and embraces us is the Intellect that influences us, and forms the link between us and God.

The Assimilation of Sacred Ideas

Finally, always remembering that by "intellect" Maimonides means a human faculty far different from mere conceptual activity, we are told that the true perfection of man involves the inner absorption of ideas. So crucial is this that upon it hinges the attainment by individual men of the property of immortality:

> The fourth kind of perfection is the true perfection of man; the possession of the highest intellectual faculties; the possession of such notions which lead to true metaphysical opinions as regards God. With this perfection man has obtained his final object; it gives him true human perfection; it remains to him alone; it gives him immortality, and on its account he is called man. Examine the first three kinds of perfections [wealth, physical health, morality], you will find that, if you possess them, they are not your property, but the property of others. . . But the last kind of perfection is exclusively yours; no one else owns any part of it, 'They shall be only thine own, and not strangers' with thee' (Prov. v. 17) . Your aim must therefore be to attain this [fourth] perfection that is exclusively yours, and you ought not to continue to work and weary yourself for that which belongs to others[4]. . .

I will not presume to rephrase these remarkable passages, nor attempt to analyse what Maimonides means by the power of Providence. All that we can confidently say is that what we call "attention" is the tip of a great iceberg, and that its development is the key to man's coming under the higher influences both in the universe and in himself. Power, which is the ability to live in the world of real causes, begins with the growth of human attention.

The Need for Active Attention

The traditions teach us that man loses everything unless he is able to listen, to see, to be present both to the lower and the higher elements in himself. In these traditional formulations, man is a bridge; and the bridge is awareness—awareness that is evoked by the struggle for active attention. Apart from that, he is naked, powerless.

This must surely be a central meaning of the story of Adam and Eve.

Adam, the active aspect of man, is commanded by God to stay by the side of Eve, the passive aspect (to "cleave unto his wife"). But, separated from Adam, Eve is beguiled, tricked, by the serpent and judges the apple only by its appearance. What was meant to be active in man has submitted to what was meant to be passive ("thou hast hearkened unto the voice of thy wife").

Is failure of attention the original sin?

Without active attention, is it ever possible for us to see the inner aspects of reality? Is it because of passive attention that we are beguiled by appearances, both with regard to the nature of the universe and the teachings which are offered to us?

Is it because of failure of attention that desire shapes our thought and understanding, and therefore our action? To be without real power: Does that not mean to act in a false world, a world that is a construct of the ordinary, passively attracted mind?

NOTES

1 Titus Burckhardt, Alchemy, Penguin Books, Inc., 1972, pp. 23—27.

2 Vivekananda, "Maya and Freedom" in Complete Works of Swami Vivekananda, vol. II, Advaita Ashram, Calcutta, 1963, pp. 120—121.

3 Sermon on the Four Intent Contemplations, in Henry Clarke Warren, Buddhism in Translations, New York, Atheneum, 1963, pp. 353—76.

4 Moses Maimonides, The Guide for the Perplexed, translated by M. Friedländer, New York, Dover Publications, Inc., 1956, pp. 386—89, 395.

CONCLUSION

Science and the
Humanization of Truth

The Existence of the Path

*T*hroughout history, special names have existed for systems of ideas that serve the work of exposing and then unifying man's fragmented and conflicted inner nature. But familiar translations of these names like "wisdom" or "higher knowledge" no longer convey the precise power of such ideas as guides to the discipline of self-interrogation.

This discipline, taken in its widest sense to include both ideas and psychophysical exercises, also has its special name. It is called the path or the way, or sometimes "the sacred science." And as for the form of social order, the community, that pursues this Path, it too has had a specific designation, whose meaning still echoes faintly when we speak of a "brotherhood" of seekers.

The community of the earliest Christians was perhaps such a brotherhood, as was, so it is said, the original Buddhist *sangha* and the community of the Essenes around the time of Jesus. There are those who even claim that whole civilizations once existed that were structured under the hidden guidance of a community devoted to the science of self-investigation, with all that implies in terms of rule by the wise and of social forms and rituals embodying great psychological understanding. Those who make such claims often cite certain periods of ancient Egyptian civilization or the nation of Tibet before the modern era as examples.

For most of us, such evaluations of the theocratic structure of archaic civilizations must remain purely in the realm of speculation. But that need not close us to the thought that throughout all periods of history in one place or another the Path has existed, even though hidden from view. Archeological or scholarly evidence in this matter cannot be the main court of judgment. For in order to make any judgment about the existence of a school of great spiritual knowledge, it is necessary to know how to distinguish knowledge that serves the quest for self-transformation from knowledge that serves the desires and fears of ordinary, biosocial man. And few among us, scholar, scientist or otherwise, can claim to have the sensitivity that is required in order to reliably discriminate between these two kinds of knowledge.

A Sufi master living in fourteenth-century India writes to his pupils of the difference between *Shariat* (Religion) and *Tariqat* (the Path):

> The first step is *Religion* (Shariat) . When the disciple has *fully* paid the demand of Religion, and aspires to go beyond, the Path *(Tariqat)* appears before him. It is the way to the Heart. When he has fully observed the conditions of the Path, and aspires to soar higher, the veils of the Heart are rent, and Truth *(Haqiqat)* shines therein. It is the way to the Soul, and the Goal of the Seeker . . . Religion is for the desire-nature; the Path, for the Heart; Truth for the Soul. . . .
>
> *Religion* is a way laid down by a Prophet for his followers with the help of God . . . Religion consists of a series of injunctions and prohibitions, and deals with monotheism, bodily purification, prayers, fasts, pilgrimages, the holy war, charity, and so on. The Path . . . consists in seeking the essence of the forms dealt with by religion, investigating them, purifying the heart, and cleansing the moral nature of impurities. . . . Religion deals with external conduct and bodily purification; the Path deals with inner purification.[†]

Shariat (religion) therefore concerns the external life of biosocial man and the external order of society. It is said that without *religion* the life of mankind would degenerate to the level of the animal. That is to say that without the teachings, forms and patterns of life provided by *religion,* the structure of civilization would be disintegrated by paranoiac fear, by the pretensions of the emotionally riven intellect, and by the chaos caused by the undirected energies peculiar to the human organism.

[†] *Letters from a Sufi Teacher,* Shaik Sharfuddin Maneri, translated by Baijnath Singh, Samuel Weiser, New York, 1974.

Conclusion: Science and the Humanization of Truth

The implications of this distinction are vast and all-encompassing. It may be that almost everything we know of as "spirituality" is actually part of *religion* in this sense. No matter what tradition we are speaking of—whether it be Eastern or Western—most of the forms, prescriptions and ways of life that we generally know of are intended for the external well-being of biosocial man. This includes forms and laws concerning family structure, sexual relationships, dietary and hygienic rules as well as patterns of prayer and worship in all their collective and private aspects. But not only this. Also included in *religion* are ideas—the knowledge, the teachings about man and the universe!

As I understand this, it does not necessarily mean that the Path makes use of entirely different ideas, symbols, sacred writings and patterns of living—though this indeed may often be the case. The main point is that only within the precisely guided conditions of the Path can these elements function as supports in the struggle for psychological evolution. But among fragmented individuals like ourselves whose desire to awaken a new consciousness is weak and transitory at best, and most often simply nonexistent, that is to say, among the great mass of mankind, these elements of the Path have historically been given out couched in the form of *religion* to function mainly as a stabilizing influence, neither more nor less.

Seen in the light of this distinction between *religion* and the Path, we modern men and women are still in the same historical period as were the men and women of the Renaissance and the founders of the scientific revolution. We may call this period, quite bluntly, the era of the disintegration of the Christian *religion*, always remembering that by *religion* we mean an encompassing structure of publicly available ideas, patterns of family life, ritual, symbol and education that maintains the general stability of civilized life. For some five hundred years, then, the Christian *religion* has been breaking down and Western man has been turning instead to ideas that may have been meant to remain within the disciplines of the Path. It has been one of my aims in this book to ask what the consequences are of making use of such ideas without the psychological preparation they demand.

I have been asking, in short, whether modern science and all that it is bringing us is a result of the inevitable exploitation of esoteric thought that takes place during the breakdown of a culture's *religion*. If so, then the present turning toward the teachings of the East, which on the surface seems to be a movement against or beyond science, may actually also be part of the same process by which science itself arose and eventually bred elements that now threaten the life of man on earth.

Ideas and psychological methods are entering our culture which until now have been hidden by barriers of language and geography and by the absence of our own desire for new teachings about the cosmos. At the same time, in all of the sciences new facts are emerging which seriously challenge the materialistic view of the universe that has been the comfortable heritage of the scientific revolution. With one hand full of new scientific facts about the physical universe and the processes of life; and with the other hand full of powerful ideas (ancient in origin, but new to us) about the cosmic order and the human mind, what will we do? If we bring our hands together prematurely, what explosions will ensue in the form of a new and, for humanity, final round of exploiting sacred teachings? Yet if we delay even a moment too long or if we empty one hand in order to grab for more of what is in the other, what will become of us as the universal purposes of nature move to balance out the injuries we have done to her? What changes must take place in us if we are to be helped and not driven even crazier by the present accelerated eruption of ancient truths and astonishing scientific discoveries?

The Longing for Unity

Thus, although we began by asking for a new way to think about science, we end in need of a new attitude toward ourselves, especially the part in us that gravitates toward mere explanations of reality. All around us, both within and outside of the sciences, there is a yearning to heal the fragmentations and divisions that separate us from nature, from each other and from God. The search is for new, unifying concepts of the universe and the social order. But can the integration we long for ever be reached through the part of ourselves whose function is to divide and categorize?

The current attraction to ideas that emanate from ancient disciplines of the Path may be identified with this wish for wholeness. In the cosmological and psychological teachings of Buddhism or Sufism, for example, we may recognize ideas that encompass, rather than exclude—universal ideas, in the light of which both the human and the natural order were once related as parts of a living whole. But we must also understand that such teachings were given out in their complete entirety primarily to guide the struggle for unity within the individual person.

It is from this point of view that we may understand the deviation that takes place when the formulations of esoteric ideas are "stolen" from the personal disciplines of the Path to be organized and promoted by individuals who are themselves in the condition of unconscious psychological fragmentation. In the early history of science we see that such ideas (as, for example, the teachings of Hermeticism in the Renaissance) are forced to serve a completely different function than that for which they were originally intended. From being guides in the private struggle for psychobiological unification, they become merely instruments of the intellect which analyzes and categorizes. They become abstract concepts, serving only to unify external facts about the material universe. They become theories, hypotheses.

In this process the original formulations themselves must have undergone tremendous alteration. Aspects of these integrative ideas which did not serve the abstractive, categorizing function were de-emphasized or rejected as perhaps superstitious or outworn. We have already cited the transition from alchemy to modern chemistry as a clear example of this process. It is not understood that two completely different uses of ideas and knowledge are at issue in this transition. Instead, it is universally believed that alchemy was a sort of primitive chemistry which required the correction and modifications of the first modern chemists.

Torn from the disciplines of the Path, esoteric ideas may function very well to provide new unifications on the level of logic for the purpose of organizing facts discovered through scientific experimentation. But analysis and explanations do not of themselves bring self-unification. The analytic functions of the mind can serve any passing emotion or striving within the human frame. This being the case, mere conceptual unification often screens the inner fragmentation of man. Theories and explanations can lure our attention so far away from the whole of ourselves that we imagine these theories contain the clue to the unification of everything, including our own selves.

What we have called pragmatism is the clearest example of the fact that concepts devised to organize the world of appearances serve only the emotions connected to the desire for safety, the fear of pain and death and the wish for personal recognition. We have spoken of this aspect of the self in our discussion of the goals of psychiatry. In many traditions these emotions are all taken together and called simply, the "desires." Speaking just from this particular context, we may restate our main point as follows: Ideas which were intended as guides in the process of harmonizing the desires with the totality of human functioning become instead instruments for the satisfaction of one or another group of these desires. Of course, much more is involved in the exploitation of esoteric ideas, but surely this alone is enough to make us question our habitual relationship to such ideas.

It needs to be stressed that for many of us the enterprise of mere conceptual unification has long since become an end in itself. So much so that even the technical application of new concepts is sometimes looked down upon. I believe we flatter ourselves when we speak of such activity as the quest for pure knowledge. The familiar distinction between "pure scientific research" and applied science, the setting off of scientists from "mere technicians" amounts to very little when held against the reality of two streams of knowledge, two uses to which universal ideas can be put. Often, the "pure" scientist is locked in the pleasures and problems of the peripheral intellect, which then unconsciously obeys all the conflicting desires of our starved emotional nature.

As is well known, the ancient Pythagoreans and the school of Plato also held mathematics and "pure" science in high esteem. But only as part of a program for the development of higher intelligence—as part of a search, a discipline in which these activities could be experienced as a foreshadowing of a kind of thought that is free from the dominance of "the desires." That was the real meaning of the phrase "pure reason": reason that could reflect universal ideas by existing in the natural relationship of governance over the desires. It was understood that without this governance by the force of reason (which is perhaps better translated as "consciousness"), the normal emotional energy of man becomes inharmoniously distributed under the formation called egoism or the "false self."

I think it is important to bear this point about "pure" science in mind because of the sort of prestige it enjoyed for so long in our system of education. The division in our culture between "intellectual" and "practical" people is a direct reflection of the inharmonious division within ourselves between thought, feeling and instinct. But, according to the traditions of the Path, the overcoming of this debilitating internal divisiveness in man is extremely subtle and demanding work which must begin through choiceless self-study and not through the application of concepts to oneself.

Intelligence as Power

All the questions we have raised in this book are contained in the problem of how to situate knowledge in the center of our being so that the energy of emotion and instinct is informed by truth, that is, by ideas that reflect the universal structure of reality. For as we are—and the history of modern science clearly shows this—knowledge merely increases the activity of one part of our nature, leaving the rest of us unattended and the whole of ourselves even more disjointed and powerless than before. This is to say, for us knowledge only increases desire without increasing our power to act from the vital center of ourselves. It is as though one were to run a ship by bringing in ever more officers, charts and navigators without attending to the broken-down engines or the lazy, rebellious crew.

Many disillusioned people, particularly among the followers of the new religions, have attacked science for being a "power trip." It is now commonplace to blame the ecological crisis solely on modern science's preoccupation with mastering the forces of nature. The call is understandably for a more organic relationship of give-and-take with the environment. To support this call, ideas are brought in from other traditions, such as Taoism with its doctrine of flowing with the fundamental universal energy *(chi)*, or the teachings of the American Indian.

Yet I wonder, after all, if that is putting the problem correctly. For if we really felt the *need* for power over nature would not the main focus of our interest be our obvious inability to bring what we know "down" into our guts and feelings where the power to act is lodged? Instead of searching for new concepts (whether scientific or mystical), why are we not studying our relationship to thought in this light?

Something crucial has therefore been left out of our approach to this subject. The point is that the conceptual faculty of the mind captures all our attention. And it is this energy of attention which alone, we are told, has the power to unify mind and body. As we have seen in our discussion of magic, this force of attention means different things at different levels of internal integration. On one level it is mindfulness, a silent, nonperforming witnessing of the movements of our thought, feeling and sensation. But at another level it is power, the power to act from the living center of our being. This is called *will*.

Therefore, when we speak of great ideas as part of the totality of the discipline of the Path we are not speaking of concepts and explanations that organize the world and bind our attention. Rather we are speaking of ideas expressed in such a way as to support the diffusion throughout the total organism of an attention which can contain and harmonize the impulses proceeding from the emotional and instinctual sides of our nature. Is this internal diffusion of attention the embryo of what the traditions of the Path call the *soul*? Perhaps, but for us it is surely enough to recognize that without it there is no such thing as power either over ourselves or over nature. All attempts to humanize science must necessarily fail if they ignore this crucial point.

With this in mind it is clear why in the traditions of the Path ideas are never presented apart from a total discipline involving the body and the emotions (thus, again, the practical, instrumental function of ritual, sacred art and the various other forms which we have mentioned in our discussion of psychiatry). The internal diffusion of attention and the organization of the world are two entirely different movements within the human organism.

From the point of view of the Path, we are asked to recognize that human power and humanized knowledge can only exist when the latter movement flows directly from the former, when the world is organized as a direct *result* of the movement toward inner unity. I think that it is only when this relationship obtains that utilitarian concepts (as opposed to esoteric ideas) can serve their necessary function of stabilizing the communal life of man and dealing with the challenges of the physical environment

Just as the Path has nourished great religions which organize the existence of entire civilizations, so has it often supported and perhaps guided external sciences which enable human beings to acquire food and shelter, maintain their safety and satisfy their normal physical and social desires. But in a traditional society the sciences were never meant to satisfy the myriad contradictory desires of disharmonious man. The present overdevelopment of technology could only have taken place in ignorance of this point.

As we have suggested, when *religion* breaks down, piecemeal fragments of the Path sometimes become the basis of "new religion." But when *religion* breaks down, so too may the sciences break down that were supported by the Path. Then where do people turn for new concepts to organize the physical world which is constantly threatening and whose threats are constantly changing form? Again, they may turn to the ideas of the Path and in so doing ignore the primary function of these ideas, which is not to directly organize the world of appearances, but to serve in the creation of harmony and psychic force within oneself.

The Wall of Truth

In the course of writing this book I have come across very few criticisms of modern science that could stand up, in my own mind, against the impact of scientific fact. I encountered many brilliant metaphysical treatises exposing the assumptions of scientism. Yet I felt, and continue to feel, that one cannot study science without running into something which I can only call *a wall of truth*. In the last analysis, there is something extraordinarily honest and clear in the scientific ideal which even the most profound critics seem unable to weigh properly. Nor have I felt among the scientists themselves nor among their advocates an *exact* sense of the value of the scientific enterprise. I do not presume to set myself above these critics and scientists. But during the writing of this book and while criticizing science, I have felt a growing sense that I must understand for myself what it is that is also so *right* about the ideals of science.

I am persuaded by scientific fact. That is to say, I am persuaded by the truth of what I can perceive directly with my senses. And I see that the power of scientific thought is to organize such perceptions in a coherent manner for the sake of the human species' survival in the physical world.

Yet there is something in me—perhaps I should say, in us—which wishes for more, which turns to ideas and concepts for which there are no corresponding facts in my experience. Is this a weakness in my nature, as many supporters of scientism would say?

Let us proceed quite slowly. Surely, the fathers of modern science felt the inhumanness of ideas and teachings which could not be verified through one's own experience—teachings which were presented in forms that smothered the freedom of the mind. Thus, in the early modern era there was a great turning toward the "wall of truth" represented by the immediacy of sensory experience: observations which *I* can make, which *I* can freely assent to without the deceptions and inner violence of blind faith. But does that mean that science was born in the worship of sensory experience? Did the fathers of the scientific revolution envision that all their ideas about reality would simply be generalizations of sensory data?

I do not think so. Yet many of them must have felt that through the renewal of trust in the senses they could be rescued from the tyranny of the isolated intellect, from the fantasies and thought dreams of dogmatic metaphysics. Then what was the attraction of the ideas of the Path which exerted such influence when institutions of the Christian *religion* began to petrify? What sort of ideas about the cosmos can attract men and women who wish not to be deceived by the mind unanchored to living experience?

I say that the great discovery of modern science was that *through the senses thought was humanized.* Through participation of the body, through the checks and corrections of the bodily senses, ideas could be brought closer to the center of the human organism. There could exist assent without blind faith. But I also say that in general *this principle was never sufficiently valued, not even by the founders of modern science.*

What is this principle? To put it succinctly: Knowledge of the universe must involve the human body as an agent of knowing in harmony with the intellect.

But this is exactly the principle that separates sacred ideas from mere concepts and explanations. The teachings of a Path are so presented that they cannot be understood by a fragment of human nature, by the mind alone or by the emotions alone. Mind, feeling and body must enter into a more harmonious relationship for these ideas to be digested. At the same time, these ideas are meant to serve as guides for this process of harmonization. What does this mean in the present context? It means that we must entertain the possibility of the sensory experience of universal ideas—*the possibility that there exist finer levels of sensation within the human organism.*

We may say, then, that in part science arose in revolt against the tyranny of metaphysics without sensation, and that it continued to exert its power as a check against isolated thought associations and mere speculation. Where it failed, where it deviated, was in the assumption that in ordinary, technologically assisted sensory experience it has reached the limits of the body's possible contribution to knowledge. That is why the ideas of the Path underwent a change, a so-called "correction." From the point of view of the Path, it is not simply the intellect which science underestimates, *it is the human body as an instrument of knowledge—the* human body as a vehicle for sensations as direct as ordinary sensory experience, but far more subtle and requiring for their reception a specific degree of collected attention and self-sincerity.

The power of the ideas of the Path, their attraction, was that they could be verified by sensations of the external world. The error of science was to accept *only* that part of these ideas which could be so verified, without grasping the possibility of more subtle forms of sensation, of what is called in certain traditions the inner sensations, through which a fuller verification of the ideas of the Path is obtained.

We have come upon a key to the evaluation of modern science which has gone unnoticed by most contemporary observers. That is: Modern science rediscovered the need for cooperation of body and mind as instruments of knowledge. Having lost contact with the discipline of the Path, religion in the West had ceased to provide the experiential verification by means of which ideas enter into the being of man. The introduction of sensory verification was a first step toward returning to the human relationship to serious ideas. Through sensory experience and guided by ideas that come from a freer level of intelligence, the structure of the physical universe was glimpsed, while the vagaries and tyranny of the isolated intellect were beaten back. Yet the short-term pragmatic power of this approach to knowledge distracted modern man from deepening the discovery of the meeting between sensation and thought. Concepts which merely organized sensory data became the model of true ideas. With this was lost the possibility of a discipline through which universal ideas could be blended into the heart of man through subtler levels of sensation and feeling accessible within the framework of the Path. The real meaning of unification and the actual source of power, or will, in human nature was never seen in the mainstream of scientific thought. To acquire power, the modern era turned to the mechanisms of *thought* rather than to *consciousness*. As a result, the emotions—unharmonized, untouched by the ideas of the intellect—continued their work of unconsciously governing the life of Western man under the formation of egoism. Ideas which could have guided the harmonization of mind, feeling and instinct became instead mere explanations which divide and analyze, and through which unity can never be obtained.

Therefore, it is futile to insist that science reintroduce Mind, or God, into its world until we ourselves are able to introduce Mind into our own inner world. Futile to demand of science that it make use of sacred teachings which we ourselves do not understand because we have never carried through the labor of studying ourselves in their light.

Throughout the history of civilization the great traditions have offered human beings a door on the other side of which there stretches the long and difficult path to self-knowledge and self-transformation. But it is said of the guides who stand behind that door that their sole task is to conduct us forward; no promise is given that those who are distracted will ever find their way back again. Legend also has it that what is nectar on the far side is poison on this side. Therefore, in the past the door has been well guarded by the institutions and forms of Tradition.

What does it mean, then, that these guardians seem to have vanished in the present age?

INDEX

INDEX

Index

Index

INDEX

Index

INDEX

JACOB NEEDLEMAN

Jacob Needleman is Professor of Philosophy at San Francisco State University, former Visiting Professor at Duxx Graduate School of Business Leadership in Monterrey, Mexico, and former Director of the Center for the study of New Religions at The Graduate Theological Union in Berkeley, California. He was educated in philosophy at Harvard, Yale and the University of Freiburg, Germany. He has also served as Research Associate at the Rockefeller Institute for Medical Research, as a Research Fellow at Union Theological Seminary, as Adjunct Professor of Medical Ethics and the University of California Medical School and as guest Professor of Religious Studies at the Sorbonne, Paris (1992). He is the author of *The New Religions*, a pioneering study of the new American spirituality, *A Little Book on Love, Money and the Meaning of Life, Lost Christianity, The Heart of Philosophy, The Way of the Physician, Time and the Soul*, and *Sorcerers*, a novel, and was General Editor of the Penguin Metaphysical Library, a highly acclaimed selection of sixteen reprinted texts dealing with the contemporary search for spiritual ideas and practice. He was also general editor of the Element Books series, The Spirit of Philosophy—aimed at re-positioning the teachings of the great philosophers of the West to show their relevance to the modern spiritual quest. Among the other books he has authored or edited: *The Tao Te Ching* (Introductory Essay); *Gurdjieff: Essays and Reflections on the Man and His teaching; Consciousness and Tradition, Real Philosophy, Modern Esoteric Spirituality* and many others. In addition to his teaching and writing, he serves as a consultant in the fields of psychology, education, medical ethics, philanthropy and business, and is increasingly well known as an organizer and moderator of conferences in these fields. He has also been featured on Bill Moyers' acclaimed PBS series, "A World of Ideas." His most recent book *The American Soul: Rediscovering the Wisdom of the Founders*, was published by Tarcher/Putnam in February of 2002.